JAR Private Pilot Studies

Phil Croucher

About the Author

Phil Croucher holds UK and Canadian professional licences for aeroplanes and helicopters and around 7000 hours on 35 types of aircraft, with a considerable operational background and training experience from the computer industry, in which he is equally well qualified. He has at various times been a Chief Pilot, Ops Manager, Training Captain and Type Rating Examiner for several companies. He can be contacted at **www.electrocution.com**

Table of Contents

Introduction

Welcome to the world of flying! Many of you will be studying the material inside these pages because you eventually want to fly for a living – others will only be interested in a rewarding hobby. Either way, you must start somewhere, and this is it!

As with anything new, a lot is very strange at first, but you will soon get used to the strange noises and begin to settle down. Once that happens, you will find that flying demands a good deal of concentration – it is not something to be approached lightly, although there's no reason why it can't be fun. There are risks involved with almost everything we do, it's part of the price of living on a planet but, with aviation, you can't fix things once you are up in the air, which is why there is a lot of preparation before you get airborne, to ensure that as little as possible goes wrong or, if it does, you are better able to cope with it. It's safe to say that at least as much work on the ground should go into every flight hour.

The Private Pilot's Licence has pretty much the same syllabus throughout the world, because most countries adopt the ICAO standards. JAR, of course, is different, and can get a bit more technical. Having said that, the subject matter is not demanding, but it all needs to be learnt at the same time, as opposed to doing each subject by itself. It can also require a lot of memory work, especially law, but even that has been simplified in this book as much as possible, and translated into Plain English. A little knowledge of physics also helps, but all you need is included here.

Kind regards

Phil Croucher

www.electrocution.com

Principles of Flight

Why does anything fly in the first place? You will find out in this chapter, which covers aeroplanes first, and helicopters later, because they share many of the basics and it saves me typing it all twice.

Definitions

All the following are explained further in the text:

- *Inertia* - the tendency of a body to remain at rest, or at least carry on with what it's doing - in other words, not to change its present state.

- *Momentum* – the quantity of motion in a body, or a tendency to keep right on going.

- *Equilibrium* - a state of balance between forces.

- *Centrifugal Force* acts outwards along a radius of a curve.

- *Centripetal Force* acts inwards along a radius of a curve.

- *Acceleration* – the rate of change of motion in speed and/or direction.

- *Nose* – the front part of the aircraft, where the cockpit is.

- *Tail* – the rearmost part, with the rudder assembly and horizontal stabiliser. Sometimes known as the *empennage*.

- *Wings* – the lift producing surfaces, traditionally forward of the tailplane, but a *canard* has them the other way round. In a helicopter, the rotor blades are the equivalent.

- *Ailerons* – moveable surfaces on the rear of the wings, at the outer ends, that alter the way the wing produces lift, in order to make it go up or down, most useful when turning.

- *Stall* – a condition of flight where the wings stop producing lift and the aircraft is no longer able to stay airborne, when the machine is not going fast enough for the conditions.

- *Angle of Attack* – the angle formed between the wing and its path through the air.

The *airframe* is the complete structure of an aircraft, without the engines and instruments. It will be as light and as strong as possible, because many forces are encountered in flight, like *compression, tension, torsion, shearing* and *bending*, so stuff like wood, fabric, aluminium or carbon fibre are used (aluminium is too soft by itself, so it will be mixed with copper, manganese or magnesium for strength). Instead of being solid, where bulk is needed, a *honeycomb construction* will keep things light. This is a framework of short hexagonal tubes covered over both open ends by metal sheeting.

The *fuselage* is where the pilot, passengers and cargo are placed, and to which any wings, tailplanes, tailbooms and main rotors are attached. Older aircraft will be made of a *truss construction*, or *frame and skin*, where aluminium or steel tubing is joined in a series of triangular shapes (like the tail boom of the Bell 47 helicopter), then covered with metal or fabric (in this case, the metal acts merely as a cover, making no contribution towards strength). One disadvantage is that cross bracing takes up a lot of space.

More modern machines use *monocoque*, which is a development of *stressed skin*, where the outside covering itself is rigid and takes the stresses of flight, and supporting devices inside, like *formers* held together by *stringers*, or *longerons*, provide the shaping (this is also a

typical method used on the tail booms of most modern helicopters).

Formers give the fuselage its basic shape, and are assembled one after the other, changing in size as required. They will absorb *torsion* and *bending* loads. *Longerons* run fore and aft, keeping the formers together, spreading the load between them and stiffening the structure in general. *Bulkheads* are similar to formers, but tend to be found at both ends of the fuselage, at each end of a compartment, or when more strength is required.

A *firewall* is a fireproof partition that separates the engine compartment from the cabin. It is normally made of stainless steel.

In a wing, *ribs* are the equivalent of formers, and they are held in place with *spars*, which perform a similar function to longerons. Modern wings will also contain fuel tanks, which may or may not contribute to the strength of the wing.

An egg is a good example of a monocoque structure, which is handy, as *coque* is French for *eggshell*. Aside from saving weight, the big advantage of monocoque is that it leaves more space inside. Older flying boats, made of wood, were among the earliest examples.

Airflow

Air is a liquid medium, meaning that it behaves rather like water (as is shown by submerged aircraft, which will "fly" to the bottom of the sea, miles away from where they splash down). It is also compressible, and can flow and change its shape.

The speed at which an object moves through the air is called the *airspeed*, and it doesn't matter whether the wind flows over it, or the object itself moves – the effects are the same. Up to a certain critical airspeed, airflow round a body is quite well-behaved, after which it breaks up to form vortices that may interfere with any lifting action.

The Aerofoil (or Airfoil)

This is the official name for a wing, or any other device that creates a lift reaction out of thin air (in order to get airborne in the first place, the *lift* must always be more than the *weight* of the aircraft - in the cruise, of course, they should be equal).

The complete force produced by any aerofoil is called the *total reaction*, which can be split into two forces, called *lift*, which acts at right angles to the airflow, and *drag*, which acts parallel to it.

In the diagram below, the thrust and lift vectors are longer than those for their opposites, weight and drag, so you will fly forwards and upwards:

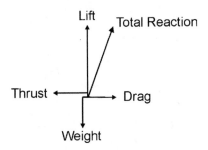

You will also notice that the lift/weight and thrust/drag vectors are offset from each other. This is to create couples around the lateral axis to produce pitching moments when

lift and thrust are taken away (as with an engine failure), placing the machine in the correct attitude. They will be balanced in normal flight by forces produced by the tailplane.

The *chord line* is the straight line joining the leading and trailing edges of an aerofoil:

The *Centre of Pressure* is the point on the chord line through which the resultant of all forces (i.e. total reaction) is said to act:

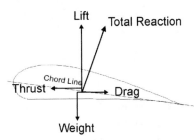

It moves forward steadily as the angle of attack increases (see below), until *just before* the *stalling angle*, when it moves rapidly backwards, creating such a long couple between it and the Centre of Gravity that the nose pitches forward. The range can be as much as 25% of the chord length.

The stalling angle is that *above* which the aerofoil stalls. It is where lift is at its maximum. Although lift is still being produced above it, it is not enough to support the aircraft.

The C of G is an imaginary point around which the aircraft is balanced, and is normally forward of the C of P anyway. If it is too far forward, the couple will be long enough to produce a large nose down pitch from the lift/weight vectors. There will then be a longer distance between the C of G and the elevator, which will tend to make the machine longitudinally overstable (see *Stability*, below), meaning that you will need more control input to pull the column back on landing, and you may run out of range.

To start off with, a wing is placed at an angle on the airframe called the *angle of incidence*, which is purely a figure out of the designer's head, although there are advantages in having it as small as possible, in that you can improve visibility and reduce drag in the cruise because the nose will not be so high (in practice, it is set at the best *lift/drag ratio*, or the point when you get the most lift for the least drag). This angle may vary throughout the length of the wing, being maximum at the root and minimum at the end, in a process called *washout* (or *washin* if you go the other way. The difference is that the former decreases lift and the latter increases it). The angle of incidence changes this way because the outer edges of the wing (or propeller, which acts on the same principle) will be moving faster than the rest in some manoeuvres (a turn, for example), creating more lift and stress. In addition, washout allows the outer parts of the wing to still be creating lift at slower speeds when the inner edges are stalled, as they might be when landing. You can get a similar effect by changing the *shape* of the wing from root to tip.

Later, in flight, the wing will make another angle with the *relative airflow*, or *relative wind*, which will be the *angle*

of attack. Te relative airflow is just the direction of the air that keeps the aircraft up, which goes the opposite way to the flight path. In other words, it is the direction of the air relative to the wing, irrespective of whether the wing is pushed through the air or *vice versa*, or a combination. Relative wind has nothing to do with the real wind.

The *angle of attack* is the angle at which the wing meets the air, or, more technically, at which the *chord line* (that joins the leading edge with the trailing edge) meets the *relative airflow* or the flight path — do not confuse it with the *angle of incidence*, mentioned above. You can either fly at a high speed with a small angle of attack, or a slow speed with a high one, up to the accepted maximum of around 15° - as the angle of attack increases, there is more frontage to the airflow, increasing *drag* markedly.

Put simply, the airflow will hit the underside of the wing, to be forced downwards, forcing the wing up.

This is a bit of a brute force solution, so the wing will also be shaped to help things along with the *venturi effect*, discovered by Bernoulli, a principle made use of in carburettors, described in the *Airframes, Engines & Systems* chapter, and air-driven instruments (see *Instruments*). Bernoulli found that the pressure of a fluid decreases where its speed increases or, in other words, in the streamline flow of an ideal fluid, the quantity of energy remains constant - there is a given amount of energy involving speed and pressure, and each affects the other directly.

If you take a tube with a smaller diameter at its centre than at either end, and blow air through it, the pressure in the centre is less because speed and pressure interact with each other, in that, if you increase one, the other decreases:

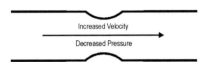

Increased Velocity

Decreased Pressure

In this case, the air being forced around the obstructions in the middle has to increase speed to keep up with the rest because it is taking a longer path. Of course, the same molecules of air don't meet up at the other end – those taking the longer route may be up to 30% of the distance away, depending on the angle of attack. Since speed is increased, pressure is reduced. This system also pulls fuel into a carburettor, and it's also the reason why a door closes by itself if left slightly ajar – there is less pressure in the gap between it and the door frame as the air moves through it.

If you take the top half of the tube away, the phenomenon still works on the remaining half, which looks like the top surface of an aerofoil. You can see this yourself by taking a large piece of paper and folding it back over the top of your hand, keeping hold of it with your fingers. If you blow across the top, you will see the paper rise. Used sideways, this is how yachts make use of the wind to get along.

The aerofoil will therefore have a natural tendency to go up or, looked at another way, to pull air down, to

the low pressure area on the top and help the brute force effect. Around two thirds of the total lift comes from the reduced pressure effect across the top, not forgetting the higher pressure underneath.

The *wing loading* is the average weight lifted by each square foot of the wing. It works in a similar way to loadspreading, in that a larger area has a lesser loading.

Aeroplanes

Aeroplanes are either *low wing*, like most Piper aircraft (see overleaf), or *high wing*, like the average Cessna:

High wing planes may have *struts* to help keep the wings up, as well as internal bracing. Both types are also called *cantilever*, or *semi-cantilever*, respectively.

Aeroplanes are *single*- or *multi-engined*, with *fixed* or *retractable* undercarriages, if they are landplanes, skis if they land on snow, and floats if they operate on water:

A *monoplane* has one pair of wings, while a *biplane* has two (the Red Baron's *triplane* had three. I forget what Snoopy had). The shape of a wing as viewed from above is known as a *planform*, and could be *rectangular*, *tapered* (from root to tip), *elliptical*, *delta* or *sweptback*. Large, wide ones, for example, are good for large transport aircraft, and short, stubby ones will be found on fast sports aircraft. The *aspect ratio* of a wing is the relationship between its length and width, or *span* and *chord*. You could have two wings of equal surface area but different aspect ratios, depending on what they were designed for. The higher the ratio (i.e. the longer the wing relative to its width), the more lift you get, with less induced drag and downwash (as with gliders):

However, it does stall at a lower angle.

An aeroplane's rated strength is a measure of the load the wings can carry without being damaged. Light aircraft can take total loads in three categories:

- *Normal*, 3.8 x the gross weight

- *Utility*, 4.4 x gross weight

- *Acrobatic*, 6 x gross weight

Naturally, there is a safety factor involved, but the above should not be exceeded.

Normal or utility categories do not allow manoeuvres with high positive and negative load factors. Bank angles would normally be inside 60°.

Flight Controls

When airflow over them is high, the controls have a positive feel. They are less responsive at slow speeds, which is a point to remember when flying low and slow (also, because of the nose high attitude, the control surfaces may not have any airflow over them at all, which is why some aircraft have the tail at the top of the fin). In fact, the effectiveness of any control depends on its distance from the Centre of Gravity, the size of the control surface, its speed through the air and the degree of movement.

Control surfaces in small aircraft are usually activated by cables and pulleys, or rods and tubes.

The *elevator*, *rudder* and *ailerons* are attached by hinges to the tailplane (or horizontal stabiliser), fin and wing trailing edges, respectively (a *Canard* is a horizontal stabiliser mounted on the *front*, with the advantage of a longer moment arm, so they can be smaller).

The *elevator* controls pitching by increasing the angle of attack above or below the tailplane, according to whether it is raised up or forced down by movement of the control column in the cockpit (if the column is pushed forward, the elevator is

forced down into the airflow
underneath the stabiliser, the angle
of attack is increased, the tail rises
because more lift is created and the
nose goes down, the opposite if
pulled back). Sometimes, there is no
elevator, but the whole stabiliser is
moved, in which case it is a *stabilator*.

The *rudder* does much the same
thing, only sideways, making the
nose *yaw*, or move left and right. It is
controlled by the foot pedals –
whichever one goes forward moves
the rudder to that side, where more
lift is created and the fin is forced
sideways in the opposite direction, to
produce a flat turn with a skid (you
don't use the rudder to turn, but to
fine tune one initiated by the
ailerons, or stop it going the wrong
way – see *Low and Slow*, below).

The *ailerons* make the aircraft *roll*
around the nose. If you move the
control column to the left, the right
aileron goes down, increasing the
angle of attack on that side, and the
left one goes up, decreasing it,
causing a roll in the same direction.
To counteract *Aileron Drag*, which
comes from the downgoing aileron,
you might see *frise* or *differential*
ailerons used. With the former, the
downgoing aileron is streamlined.
The latter moves the down aileron
through a smaller angle.

In fact, the frise aileron's hinge is
offset, so a portion of the leading
edge of the downgoing aileron sticks
out into the airflow, to create a little
drag to reduce adverse yaw. It also
produces a slot (i.e. a gap between it
and the rear of the wing) to smooth
out the airflow over it:

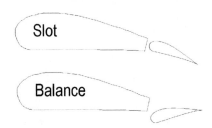

The controls will move the aircraft
in one of three axes – *pitching* (nose
up or down), *rolling* (wings up or
down) or *yawing* (nose left or right).
They do not move in isolation,
however – an adjustment in one
causes a secondary effect in another
and must be allowed for, as we shall
see in the discussions below. For
example, an uncontrolled yaw
eventually results in a roll, because
one wing will be moving faster and
will generate more lift on that side.

Your instructor will need to
demonstrate certain manoeuvres,
and there is a well established system
of establishing who has control of
the aircraft at any given time. The
person handing over the controls
will say "You have control", while
the person receiving them replies
with "I have control".

Trim
Depending on the net result of
power and control positions, it may
take more physical force to keep the
aircraft in a particular attitude. That
is to say, for any combination of
power and control position, they will
move freely with a certain range, but
take a lot of force to go outside of it.

These extra forces can be *trimmed out*
with a wheel or similar device which
operates a very small control surface
in the elevator (for example), so you

have a control surface within a control surface. The wheel moves the surface up or down in the airflow, which moves the elevator the opposite way and does the work you would otherwise have to do to keep it there. If the trim wheel is moved forward, it forces the trim surface upwards, which creates more lift between it and the elevator, which therefore is forced down, creating more lift underneath the tail which lifts and forces the nose down. The thing to remember for exams is that the control column, when moved forward, moves the elevator *down*, whereas the trim wheel moves its attached surface *up*.

Power affects trim tabs, as more airflow varies the sensitivity of the controls. Reducing power makes the nose pitch down because the trim tab has become less effective and cannot hold the nose in position.

Trim surfaces may also be found on rudders, depending on the complexity of the machine.

You may occasionally see a *fixed trim tab*, which is there to provide a fixed amount of trim to make the machine fly true (it may be one wing low, for example, from the factory). It must only be altered by an engineer. Fixed tabs are used on helicopter rotor blades to make them fly higher and lower with respect to each other, with the goal of making them fly in line, to reduce bouncing.

An *anti-balance tab* moves in the same sense as the main surface, and is there to *increase* the force required to move the control. The further it is deflected, the greater the force (the angle of attack increases at a greater rate on the tab). This prevents over-controlling and overstressing the aircraft, especially where controls have a low aerodynamic loading.

Balance

At high speeds, control surfaces may flutter because of buffeting. To prevent this, a streamlined balancing weight (usually lead) is fitted forward of the control surface's hinge. It may be inside the control surface itself, or fitted externally (*Mass Balance*).

Sometimes, part of the control surface is placed forward of the hinge line, so that airflow hitting it will help the pilot move the controls (known as *aerodynamic balance*).

Flaps

These are hinged devices on the trailing edges of wings, inboard of the ailerons, that temporarily increase the lift producing areas for certain modes of flight, like landing, and sometimes takeoff (not in the PA 31, or aircraft without enough power to overcome the extra drag that reduces acceleration), where you might be going very much slower than normal and need a boost – in fact, flaps produce the same lift at lower speed by increasing the upper *camber* (exam question), and the negative pressure underneath because the chord line moves further down at the rear and changes the angle of attack against the relative airflow (pushing the nose down restores the original angle).

Thus, the reason for using flaps (or any other low speed lift-producing device) is to change the shape of the type of wing required for high speed flight into one suitable for low speed flight, otherwise you would need several miles of runway to get airborne. Or land, in which case you would need sturdier (and heavier) undercarriages.

However, there is a point beyond which the extra surface structure in the airflow produces more drag than lift, which is made use of when deliberately trying to bring the speed down, as with a short field landing, or increase the angle of approach without much sacrifice in speed. Sometimes, the ailerons are made to move in sympathy with flaps.

Various flap designs create different effects (all try to reduce drag), but the *Fowler Flap* is generally considered to be the most efficient – they do not just drop down from the wing, but slide out from the back:

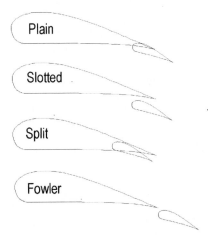

There is a *maximum flap extension speed*, as they are not designed for high speed flight. Lowering flaps will generally force the nose down, as they make the centre of pressure move backwards, but sometimes they affect the airflow over the tailplane enough to go nose-up.

Once lowered, flaps should not normally be raised until actually on

the ground. On a missed approach, they should be raised after power has been applied, in easy stages.

Other Wing Devices

Slats are small aerofoils that open forward of the main one to smooth out the airflow - when the angle of attack is high, they do this automatically (low pressure just behind the leading edge sucks them out. They are pushed back in by air pressure when the angle is low). They are usually found near the wing tips on the leading edge to help with lateral control.

Slots are openings a little bit back of the leading edge of a wing that allow high pressure air underneath the wing to pass through them at high angles of attack to the low pressure area, to re-energise it and extend the laminar flow by reducing eddies.

This reduces loss of lift and drag, and increases the stalling angle of attack. Slots may go across the whole length of the wing, or just where the ailerons are.

Manoeuvres

Taxi

Before you get to take off, you have to get from where you are to the runway, and return when you land. The "roads" that get you there are called *taxiways*, and they are usually identified with letters, such as *Taxiway Alpha*, or *Taxiway Bravo*, etc. You will be told the sequence to use when you first get clearance from ATC to start moving. The clearance may involve crossing a runway or two to get to the one you want – you should still check to see that nobody is using them! Clearance to *enter* your runway is entirely separate – taxi clearance is only to get you there.

To get started, you may need what seems like a lot of power, but this will reduce to a very small amount once the wheels start rolling. The first thing you do then is test the brakes with a small dab on the toe pedals, and bring the machine to a stop again.

Once under way, the machine will want to head into wind – it is only held straight by friction from the tyres, but you can help by positioning the controls. Moving the control column into the wind direction deflects the ailerons so that they help control direction. A high wing aircraft with the wind coming from one of the rear quarters (SW, SE, etc.) should also have its elevators down (i.e. control column forward), but this is also true to a lesser extent for other aircraft. On a rough surface, pull the control column back.

It is usual to initiate a turn to the left and right, if there is no opportunity to do so anyway, to check the instruments. The artificial horizon should always be level, and the compass and DGI increasing and decreasing according to the turn. The needle and ball (or turn coordinator) should also indicate the correct direction.

Try not to use the brakes to turn – you should be able to use the rudder pedals by themselves. Wearing out

the brakes costs money, and you don't want to do this unnecessarily. Use them smoothly at all times. It is considered to be bad practice to use high power to taxi, and use the brakes to slow you down. A brisk walking pace is recommended. If you do use a brake to turn, the inside wheel should not be stationary, or the tyres will wear.

Takeoff

You should normally do this into wind as much as possible, although circumstances sometimes dictate otherwise – check your Flight Manual for the maximum crosswind limits your machine can take, but most are certificated with the ability to handle 90° of crosswind at 20% of the stalling speed.

Line up straight on the runway, and check that the compass and DGI show the right numbers. You should not experience any yaw to the left in a nosewheel aircraft unless the nose is lifted, as it would be for a soft field takeoff (it will reduce slightly once the tail comes off the ground in a taildragger). Keep full power on until at least 500 feet off the ground, at best rate of climb speed, unless you have obstacles, in which case use best angle speed.

In a crosswind, position the ailerons as if you were turning into wind, which will stop the wind wing rising. You can reduce this as speed is gained. Once airborne, you must stay airborne (because of the sideways movement if you settle again), and you must come off cleanly in the first place. Then make a coordinated turn into wind, until you have the right heading for the drift.

Where space is limited, you have to accelerate as quickly as possible and configure the aircraft for a takeoff at slow speed, which means using some degree of flap (check the Flight Manual). to get maximum acceleration, you need maximum power without movement, which means both feet on the brakes. Assuming the power is what the Flight Manual says you should expect, release the brakes gently and raise the nose *when you reach the correct liftoff speed.*

Another way (used by bush and mountain pilots) is to keep the flaps up during the backtrack, and increase power to get as much speed as possible in the turn. Once you have speed on the takeoff roll, select full flaps and away you go. Full flaps create drag, so you want them up as soon as possible – you can do this *very* slowly while still in ground effect, keeping the nose down.

Soft and rough surfaces need more distance, and the idea is to do as much as possible on solid ground so you can get off without stopping, otherwise you might get stuck. You need the nose high, or at least the weight off the nosewheel, so your liftoff speed will be slower than normal, close to the stall. Fly level with wheels just off the ground until you have the speed for climbing.

When trying to get over obstacles, be prepared for a change in the wind that may stop you getting over them.

The Circuit

This is not just a way of making sure that everyone follows the same one-way route around an airfield, but a good exercise in precision flying.

The existence of the circuit is the reason why airfields have to be avoided by a minimum distance.

The traffic circuit is rectangular, usually left-handed, and consists of 5 parts – the *takeoff, crosswind leg, downwind leg, base leg* and *landing.*

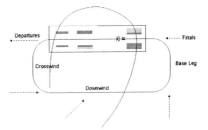

The takeoff phase lasts until 500 feet, where you make a 90° turn crosswind to level off at circuit height (usually 1000 feet, but check, in case they surprise you). Another 90° turn takes you downwind, and you report being on the downwind leg when approximately abeam the tower. You also do your pre-landing checks, leaving the landing gear for the base leg, which is another 90° turn at the end of downwind.

While in the circuit, keep a good lookout and be *very* aware of your position relative to other aircraft, especially at Biggin Hill! Adjust your spacing as necessary. Use the downwind leg to plan your final approach, according to conditions.

You generally **leave the circuit** either straight ahead, or off the crosswind or downwind legs (don't climb until clear). You **join** on the downwind or base legs, or maybe directly on finals on a slow day. You can also join at 45° to downwind, or from the overhead, merging into the downwind leg from above, having made a descending left -hand turn.

DO NOT TRY TO TURN BACK TOWARDS THE FIELD IF THE ENGINE FAILS IN THE TAKEOFF PHASE – LAND STRAIGHT AHEAD.

READ THAT AGAIN.

AND AGAIN.

Straight and Level

This is the basis of all other flight attitudes – many pilots regard doing it well as a matter of professional pride. This makes sense, because it is where you will spend most of your flying life.

Correct small amounts of yaw (less than 10°) with rudder only – larger amounts need aileron as well to prevent rolling. In a similar vein, small changes in altitude are less than 100 feet, for which you can just use the elevator. Larger amounts need power as well.

After initiating a manoeuvre, it will continue unless you centralise the controls, so, having started a turn, for example, return the controls to where they started when you get to the position you want.

Turns

A "proper" turn is one in which the aircraft is in balance, and there are no acceleration forces, unless you are climbing and descending deliberately. That is, there is a constant rate of change of direction, maintaining height, and the forces acting on the machine are in equilibrium. You turn by making the aircraft bank in the required direction with ailerons.

Having done that, some of the lift force is applied to the turn, so the lift vector is reduced. You must therefore apply some backwards control column movement to force the tail down and the nose up, at the cost of a little speed (if you want to keep the same speed, you must also apply some power). There will also be a yawing moment applied the opposite way to the turn, because the upgoing wing has more aileron in the airflow, and is producing more drag to slew the nose round (*adverse aileron yaw*). This will need a little rudder movement to keep the ship straight, although this can be allowed for in modern designs by adjusting the movement of the ailerons, or the shape, like the *Frise aileron*, which produces a counter drag from a lip that appears underneath the wing when it is moved upwards. You will also need a slight force to stop the machine turning, because the outside wing will be moving faster and producing more lift.

The greater the rate of turn, the more the lift must be increased to maintain height, and the more the weight *artificially* increases to keep the forces balanced, which is one good reason for not being overweight, because you never know when you will need the power. At 3G, for example, the weight is the equivalent of *three times* what it would be in straight and level flight.

A turn will continue unless the controls are centralised. The nose drops because the lift vector moves away from its opposition to gravity (weight). You can either increase speed or the angle of attack, but the latter is preferable, although power is required for steep turns (see below). To judge the right angle of attack, you must get used to the position of part of the aircraft against the horizon (usually the nose or instrument panel), which will be different in a left or right turn. In practice, as long as you keep that position, you will stay level during the turn, only checking the alitmeter occasionally to make sure, because you should keep the lookout going.

Remember that a bigger angle of attack means more drag, that requires a bigger angle, and so on.

Centrifugal force will tend to alter the longitudinal axis of the aircraft against the arc of the curve. Look at the turn coordinator. A slip or skid will alter the wing's lift-producing characteristics and force you into unwanted adjustments.

You need to start rolling out slightly before the point at which you wish to end up, otherwise you will overshoot – you will be turning (at an increasingly lesser rate) all the time you are rolling out.

To roll out of a turn on a selected heading, lead by half the angle of bank, that is, for a 30° bank, roll out 15° before the desired heading. Use small angles of bank for small heading changes. Usually half the number of degrees of heading change is enough.

The approximate angle of bank to produce a rate one turn may be calculated with: (KIAS divided by 10) + 7 = bank angle. Add 5 instead of 7 for statute miles per hour.

Use ailerons and rudder, and relax the back pressure, otherwise the new lift vector generated in the turn will still be active and you will climb.

Climbing turns should be no more than 15°. The load factor increases with bank angle, as does the stalling speed. Also, lift is lost with bank, reducing the climb capability. The right turn will be less severe, due to torque and slipstream effects.

You can use greater angles of bank when descending (up to 30°) because the load factor is less than it would be in the climb. However nose will drop more as you roll in, and the nose attitude will be different as it will be lower at the start. Avoid a steep turn and a dive, or you will end up in a spiral dive.

Lessons involving **steep turns** are more about how to recover from them than to perform them, since they can easily turn into spiral dives, although they should be done as well as possible, since they are good for co-ordination practice. Steep turns are officially over 30° of bank, but they are much harder to do between 45-60°. A "normal" one, as an exercise, is 360°, rolling out on a specific heading (or landmark), maintaining altitude. However, there are many other reasons for steep turns, such as avoiding traffic (where you don't go all the way round) or getting out of a valley with bad weather in it, where you just want to go back the way you came as quickly as possible. A combination can also be used to slip through the only gap in a layer of cloud, where you want to remain VMC, so you will be descending as well.

Get into a good cruise, as an extra speed margin is useful for helping the machine cope with the extra loads imposed on it. Do a HASELL check (see below).

Roll into a turn and add power as you go through about 30° (climb RPM or higher). This will help with the extra load and drag factors.

Use elevator back pressure to maintain the nose attitude against the horizon. Control the turn with aileron and rudder. Check the VSI and altimeter to see if you are keeping altitude (±100'), but this is very much a secondary check compared to the nose position. Do not use back pressure alone to stop the nose pitching down, as this will just tighten the turn – use the ailerons and rudder as well, to reduce the bank angle a bit.

When rolling out, anticipate your desired heading by about 20°. The ideal way is to use the ailerons, relax the back pressure and reduce power at the same time, but in the early stages, power can be left till last. Be prepared to adjust the aircraft for up to 30 seconds afterwards, so it can settle down.

Spiral Dives
These can be entered into from turns, or spins. The turns can be inadvertent, as when trying to fly on instruments without any experience, or flying low and slow without proper training.

If the nose drops in a steep turn, so you are descending in a turn with high power, you will be very near the start of a spiral dive.

Although you shouldn't have your head too much in the office, it's actually your instruments that will give you the best clue as to what is happening. Your speed will be increasing, and altitude falling, both at increasing rates. Engine and propeller noise will increase as well.

In this situation, pulling back on the control column is the *wrong* thing to do, since it will just pull you tighter into the turn. Instead, quickly and firmly (no harsh movements!) pull the power back. Then use aileron and rudder to straighten up. Since you immediately get more lift, the nose will go up, so be prepared to relax the elevator back pressure as well (gently and carefully).

Climbing
When climbing, putting the nose up causes the speed to decrease, so you get less lift, despite the higher angle of attack. Therefore you need more power to keep going up. However, full power wastes fuel and overheats the engine – save it for emergencies, and taking off.

There are various types of climb, including *best rate*, *best angle*, *normal* and *enroute*. The first gives you the most height in a given *time*, and the second gives you the most height in a given *distance*, useful for clearing obstacles. Both, however, will make the engine run hot if used for too long, so a normal climb should be resumed as soon as possible, not only for better cooling, but because it also gives you better forward visibility to help with your lookout. An enroute climb uses a little extra power to climb at just under normal cruise speed, so you don't reduce your groundspeed too much.

The procedure to enter a climb is to change *Attitude*, *Power* and *Trim*, in that order (it's the same to level off).

In general, keep the panel or cowling slightly above the horizon, and level before the selected altitude, at 10% of the climb rate, so going up at 500 fpm means you must start to level 50 feet beforehand.

Air density (affected by height, temperature and moisture) will have an effect on your climb performance and instrument indications – this is discussed later on.

Descent
There are two types of descent, power on and power off. In the latter situation, the *recommended glide speed* in the Flight Manual will give you the maximum range. A power on descent is used for more precise control, as when aiming for a runway. To initiate descent, the procedure is to use *Power, Attitude* and *Trim*. When you reduce power, the nose pitches down, but you should not let the speed increase. There will be some yaw to the right, caused by the diminishing slipstream. To keep straight, note the attitude and any items in the windscreen with constant bearing (your landing spot).

During the descent, open the throttle occasionally, both to keep the engine warm and to clear any ice forming in the carburettor. Lowering the flaps or landing gear will steepen the angle of descent.

You use power to level, as just pulling the nose up will eventually lead to further descent (watch for left yaw – the nose will pitch up, due to the trim position). As with the climb, anticipate the level by 10% of the rate of descent.

Approach & Landing
A good landing depends on a well set-up approach, during which a lot happens, so you must be constantly aware of what's going on. The landing spot should be kept in the same relative position on the windscreen for a constant angle of approach. The idea is to land with rear wheels first on a nosewheel aircraft, and all wheels together on a taildragger.

Keep your hand on the throttle at all times, because when you need power, you will need it *now*

Somewhere between 15-30 feet, you should start the *flare*, or *roundout*, by pulling the control column back, having closed the throttle first. Keep pulling back until the aircraft settles onto the ground (when you should actually start to flare is difficult to describe, but my own preference is when I seem to be going too fast). When very close to the ground, further backward movement of the control column slows you down rather than keeps your height, and you would keep it back once a tailwheel aircraft is on the ground. You can relax the pressure with a nosewheel machine, for weight on the nosewheel to assist steering.

Crosswind landings are more difficult than crosswind takeoffs, because the controls are less responsive (there is no airflow over them from the powered prop). There are two methods of counteracting drift until the final moments. One is to keep a wing down into the wind direction (the *sideslip*, below). The other, which requires much practice, and is my own choice, is to crab in, with the nose offset into wind, and with wings level. Then the machine is straightened with rudder at the point of touchdown.

For a short field approach, leave the power on until the landing flare is completed – in other words, "drive" it on to the ground. Get the flaps up straight away. Before going in, however, bear in mind that the takeoff run will always be longer – can you get out again?

Be prepared for quick deceleration on soft or rough ground, and watch for the nosewheel – keep the weight off as long as possible. You might want to do a low approach first for inspection purposes, and to chase the sheep off.

Slipping

To get down more quickly without increasing the airspeed, you can use slipping manoeuvre, which exists when you bank, as if to turn, but you actually keep straight with rudder (this is the *forward slip*). The *sideslip* is used when landing in a crosswind, to keep straight down the approach, and kick the machine straight at the last minute.

Stalling

The stall is a condition where the wings cannot support the aircraft in the air (that is, lift is still being produced, but it is not enough).

As mentioned before, the accepted maximum angle of attack is about 15°. After that, the air flowing round the aerofoil breaks up badly, making it unable to create enough lift, as well as creating large amounts of drag. The speed at which an aircraft stalls, however, varies according to

circumstance (remember that an aircraft with a high stalling speed is *easier* to stall).

For one thing, a thick wing will produce a lower stalling speed than a thin one, as will one with a larger wing area against a smaller one. Also, as air density decreases, the stalling speed will increase until it reaches the cruise speed and you won't get any faster (although the *indicated* airspeed remains the same).

Turning has a similar effect, too, because you are artificially increasing the wing loading – the steeper the turn, the greater the increase in stalling speed.

Because of the above variables, the quoted stalling speed in the flight manual is the "clean" speed, which occurs in a straight and level glide at maximum weight with no gear and flaps down.

As an exercise, it's the point at which the nose drops down when the elevator is pulled all the way back to the stops. Since you stall the aircraft onto the ground every time you land, practice gained here can only serve to improve subsequent arrivals on the runway.

To perform the exercise in straight and level flight, do the HASELL check, place the carb air into hot, and reduce power to zero. Keep the nose level with the horizon, and be prepared for the controls to become mushy and ineffective. Do *not* use rudder, except very sparingly to keep straight, or you might end up in a spin, and you don't learn to get out of those until the next lesson. Allowing it to yaw is just as bad as using too much rudder, so don't use aileron, either, because yaw is a secondary effect of roll. In addition, aileron drag will only make the situation worse.

Keep pulling the elevator back, in attempt to maintain height. About 5-10 kts before the proper stall, you might hear the stall warning going (if you've got one), or feel a little buffeting in the controls. This is the aircraft protesting that it can't stay up in the air, and that it is reaching the critical angle of attack – turbulent air is hitting the elevator and other controls.

At the stall, the nose will pitch down, usually just after the elevator reaches its full limit of travel backwards. The dropping could be relatively mild, or quite severe, depending on the design of the aircraft (sometimes, you don't even notice it!).

The point to realise is that, in dropping the nose, the angle of attack improves enough to get lift again, and the aircraft starts flying, even if you keep the control column back, but we want to recover, so relax the back pressure, and note how much height is lost during the exercise, once you recover the cruise attitude. It will be around 350 feet, quite critical near the ground, so applying power just after relaxing the back pressure will reduce this to a minimum, sometimes down to below 100 feet.

Spinning
When spinning, the aircraft is out of control in all three axes of flight. It results from uneven stalling. A spin is basically a stall that is not straight. That is, you are turning and

descending with one wing (the downgoing one) in a permanent stall, hence the spin. The effect is a continuous roll, which causes yaw. Left to itself, the aircraft will not recover, as long as the one wing remains stalled.

You can't use aileron to get out of a spin, because of aileron drag making the condition worse, but rudder is available, so you shove in a bootful in the opposite direction to the turn, until the yaw stops. Then relax the back pressure on the elevator to pitch the nose down and reduce the angle of attack on the stalled wing (this may be done at the same time, depending on the machine). Do not use ailerons, and pull out of the dive once you have some airspeed. Then apply power as necessary.

The aircraft must stall first, so if you avoid stalling you won't get into a spin. To start a spin deliberately, shove in a bootful of rudder at the point of the stall.

Before performing any manoeuvres, however, you should do:

The HASELL Checks
Remember these well, as they are useful throughout your flying, particularly with steep turns:

- *Height* – are you high enough to perform the exercise and to recover if something should happen, without hitting the ground?

- *Airframe* – clean – flaps are in, undercarriage is up, etc.

- *Security* – hatches and harness all secure (i.e. doors closed and seat belts done up), no loose articles in the cabin that could fly about and injure you, etc.

- *Engine* – fuel is on, and enough for what you want to do, with temperatures and pressures OK, carb heat green, etc. (carb heat should be on *before* reducing power below a certain RPM, as it won't have enough power to *defrost* if any already exists).

- *Location* – no good wondering where you are when the engine stops, better find out now! Also, make sure you are not over anywhere you are not supposed to be, like congested areas, water, etc.

- *Lookout* – make sure there is no traffic above, below or around. Do a couple of turns to make sure, but *not* a steep turn at this stage, because you will learn how to do them later!

Low and Slow

This manoeuvre, which is defined as operating somewhere between stall and endurance speeds, can kill the unwary. It's commonly used on pipeline inspections, or police patrols, and is especially dangerous with steep turns. When banking hard over, your lifting aileron is fully deflected, so you can turn. Unfortunately, it's also producing maximum drag, which will tend to cause an *adverse yaw* in the *opposite* direction, that is, the aircraft wants to go right, but is being forced left, or whatever. In contrast, the aileron on the other wing has very little profile above it, so is producing very little drag. It isn't just the ailerons – the wing rolling motion doesn't help the situation, but the point is that, if

you don't use rudder to counteract this, the wing causing the yaw slows down and produces less lift, which will stop it rising as told to by the aileron. The other wing moves faster, and gets more lift. The end result is that you roll the wrong way, or at least the Wright Brothers did.

Modern design methods have reduced the risks in the normal flight envelope, but when in extreme situations, such as in a steep turn, near the stall (and don't forget that a stall can happen at any speed with the wrong angle of attack) it may well catch you by surprise if you use the ailerons too abruptly, especially when the lifting wing stalls and puts you in a spin, which is just what you don't want at 200 feet (remember the aileron's purpose is to temporarily increase the angle of attack). **Tip:** use ailerons *last* out of a steep turn. Put the control column forward and use opposite rudder first, remembering that the controls are much less effective at slower speeds.

This can also be a problem when taking off from a short strip, with both wings at a high angle of attack. Sharp movement one way or the other will increase the angle on the lifting wing and stall it the wrong way. Again, modern design, such as a twist in the wing that makes the tip ride flatter, has improved matters, but try it in a Cessna 152 (a *long* way off the ground!) to see what I mean.

Stability

The *stability* characteristics of an aeroplane describe its ability to return to its flight path after a disturbance without input from the controls. A stable aircraft is easier to fly and more pleasant, but one too much that way will not be so manoeuvreable. The *static stability* is the *initial* tendency, while the *dynamic stability* concerns the *overall* tendency, after a series of ever-decreasing oscillations – having one does not necessarily lead to the other. Its significance lies not just with you nudging controls by accident, but if you encounter turbulence, which has the most to do with knocking your aircraft off its flight path.

If *positive* stability is a tendency to return to the flight path, *negative* stability tends to move it further away in increasing movements:

You could then say that the aircraft is unstable. This could be a problem when the increasing oscillations lead you to stall or dive. Be aware that a badly placed C of G can make a previously stable aircraft unstable.

Neutral stability occurs where the oscillations are constant around the original flight path, or a new one is taken up completely.

Stability also works in the pitching, rolling and yawing planes, as in *longitudinal, lateral* and *directional.*

Longitudinal Stability
The Centre of Gravity (which is an imaginary point around which the aircraft is balanced) is designed to be ahead of the centre of pressure, to make the plane nose heavy so that, without engine power, the machine adopts the correct gliding attitude.

In the cruise, the tailplane's negative lift balances this tendency. With longitudinal stability, you will pitch when a vertical gust hits you.

Lateral Stability
This makes you roll when hit by a gust from the side. You get it if the wings are not level across their span.

The *dihedral* is the angle between the wings and the horizontal, looked at from the front – a *positive* dihedral has both wingtips higher than the roots, and enhances stability in the roll plane – if the flight path is disturbed, and the aircraft sideslips, the lower wing produces more lift to restore level flight because of the increased angle of attack. If you had your hands on the controls all the time, of course, like the Wright Brothers, you wouldn't need it.

A nhedral is the opposite, where the tips of the wings are lower in the horizontal than the roots (e.g. the Harrier, or the BAe 146):

In a high wing aircraft, the *keel effect* of the fuselage acts like a pendulum to pull it back to normal.

Directional Stability
This comes from fins, and makes you yaw when hit by a gust from the side. The fin acts like a weathercock to keep the aircraft straight – if it

yaws, the surface is struck more from the side to force the nose back.

The fin acts like a weathercock to keep the aircraft straight – if it yaws, the surface is struck more from the side to force the nose back.

One reason for the roll that results from using the rudder by itself is the *dihedral effect*, otherwise known as the *secondary effect of rudder*, if you remember your instructor's lessons, or *rolling moment due to sideslip* (the wing on the outside of the turn goes faster, produces more lift and goes up, to start the roll). Others are sweepback (above), high wings with a low C of G or a high fin and rudder, though the latter will produce a roll in the opposite direction (i.e. left with right rudder).

Dutch Roll

A combined effect of disturbing the yaw and roll axes, with more roll than yaw (if it were the other way round, it would be called *snaking*).

Essentially, the machine rolls as it yaws, arising out of sideslipping when the machine yaws. When this happens, the effective span of the wings is changed and the forward one creates more lift for a short time, because it presents more of a span to the airflow than the other one. This makes it rise, hence the roll. However, the increased lift also creates more drag to pull the wing back, starting an oscillation.

Spiral Stability

If you release the controls in a turn, the machine will either wind into the turn or come out of it by itself, indicating *negative* or *positive* spiral stability, respectively.

Forces In Flight

The four forces acting on an aerofoil are *Lift* & *Weight*, *Thrust* & *Drag*. The parts of each pair oppose each other, and must be balanced for straight and level flight. Where their points of action do not correspond, a couple is created that will affect the aircraft attitude.

Lift

Lift acts through the Centre of Pressure at 90° to drag and the relative wind. You can increase it in 4 ways, in this order:

- Increase speed for more reaction over the wings.

- Increase the angle of attack (up to the *stalling point*).

- Increase the wing area with extra devices, such as flaps.

- Fly in denser air (that is, lower or colder).

The *centre of pressure* on an aerofoil is the point where the lift is supposed to act, which varies with the angle of attack. Its usual position is around a third of the way from the leading edge, but it moves forward as the angle of attack is increased. Its most forward point is just *before* the stalling angle. Airflow is at its maximum velocity at the CP in level flight. It moves *aft* (and towards the wing root) when flap is lowered.

The *centre of lift* is an imaginary point on the airframe where the total force of all the lift producing surfaces is said to act. Normally, it should be slightly behind the centre of gravity, which is equally imaginary, and where the weight forces act through. The reason for this is to produce a slight couple that will tend to make the machine fly nose down, which is useful without engine power (thrust will normally make the couple ineffective, as will application of the elevator when thrust is reduced for low speeds).

You will not be surprised to hear that there is a formula for calculating lift, which is:

$$L = C_L \tfrac{1}{2} \rho V^2 S$$

where L=Lift, C_L is the *coefficient of lift* (the product of aerofoil design and angle of attack), ρ(rho)=air density, V=TAS and S is the wing area. The coefficient will be at its maximum at, or just before, the stall, which you can see from the formula – if it (C_L) increases on one side, the other side will increase also, until lift can no longer be produced. Similarly, increasing speed (V) will increase lift.

Weight
The opposite of lift, acting through the Centre of Gravity.

Thrust
The force that makes the aircraft move through the air, and the opposite of....

Drag
It's not all plain sailing for anything forced to move through air, as a certain resistance tries to prevent it, caused by friction from air molecules as they are forced out of the way (*skin friction*). This tendency to stick is called *drag*, which both absorbs energy and produces heat, so it needs to be reduced as much as possible. Again, to fly, the *thrust* (from the engines) must always be

greater than the total drag produced by an aircraft.

If you've ever been through a car wash, and your car is still wet, you may have noticed droplets of water remaining quite still on the bonnet no matter how fast you drive. This happens on aeroplane wings as well – large specks of dust will remain on them even through a Transatlantic flight. On the other hand, there is a point at which the air will flow smoothly over any surface. The area between the two is called the *boundary layer*, which ideally should have a laminar flow (i.e. smoothly layered) but, in practice, this doesn't happen much farther back than the thickest part of the aerofoil. In most small aircraft, the boundary layer ranges from about half an inch up to the *transition point* (where the air becomes turbulent), to around three inches afterwards, where it becomes the *turbulent layer*.

Anyhow, back to drag, which is a force that tends to slow an aircraft down, acting in the opposite direction to thrust, parallel to the relative airflow. In order of priority, it can be split up into various components:

- *Induced Drag* comes from the air's reaction to the aerofoil, or is induced from the creation of lift, so it comes from lift-producing surfaces and varies with the angle of attack, so the *slower* the aircraft the more you get (more lift, more drag). It may come from wingtip vortices, for example, and is inversely proportional to the square of the velocity, that is to say, halving velocity increases induced drag by four times. It also *increases* as an aircraft pulls out of ground effect on takeoff, as the ground will interfere with vortex formation, and can be affected by the aspect ratio of the wing.

- *Parasite Drag* comes from anything moving through the air not actually creating lift, like the fuselage, undercarriage, etc. It consists of:

  - *Interference Drag*, or the result of the interaction between various components, say the wings and the fuselage. In other words, if you added the various types of drag together, you would find the result to be less than the actual total. Interference drag is the difference.

  - *Profile Drag* is made up of:

    - *Form Drag*, from the shape of any body moving through the air. It is minimised by streamlining

    - *Skin friction*, mentioned above. It's the result of the smoothness or otherwise of surfaces.

    Profile Drag is proportional to, and increases as, the square of the speed.

Aileron Drag comes from downgoing ailerons, causing a yaw in the opposite direction of the bank.

Maximum Range Speed

This gives you the most lift for the least drag, for the most economy, and the most distance for altitude lost. Flying either side of that speed will *decrease* the range when gliding.

The *Lift/Drag Ratio* comes from dividing lift by drag. The angle of attack for the best ratio varies with the design of the wing, but is around a third to a quarter of the size of the stalling angle. It never changes, but without an angle of attack indicator, you need an indirect method of guesstimating it, such as speed, which may have to be increased slightly with aircraft weight, reducing your range, as you will be using more power, and hence fuel, to attain it.

The biggest factor concerning your range will be the wind, which will reduce your groundspeed when on the nose, causing you to use more fuel. However, a slight increase in airspeed, say 5-10%, will get you there sooner with only a slight effect on fuel consumption.

Maximum Endurance Speed

This gives you the most time in the air for least amount of fuel, which is useful when waiting for the weather to clear, or when asked to hold clear of a control zone, but, in practice, it gives you little or no controllability, so there will be a *recommended endurance speed* in the flight manual, which is a few knots above.

The endurance is longer the lower you can fly (allowing for safety, of course), and turbulence and flaps will affect the speed considerably.

There is more drag with endurance speed than there is with range speed, which is higher.

Formula

Guess what? There's a formula for drag, too, which, luckily is very similar to that for lift:

$$D = C_D \tfrac{1}{2} \rho V^2 S$$

Drag will also increase with speed.

Load Factor

The total lift divided by the total weight, the ratio being 1:1 in level unaccelerated flight. In other words, the weight carried by a wing expressed in terms of "G".

When you turn, the aircraft tries to continue in a straight line, and a force is needed to point it towards the centre of the turn – this would be *Centripetal Force*, which must be generated by extra lift from the wings. In effect, the (upwards) lift vector is reduced by the same amount, which needs to be compensated for. In a 60° banked turn, therefore, the amount of lift you need is doubled, so the load factor becomes 2. This can also be increased temporarily by sharp manoeuvres or gusts and turbulence (a gust with a speed of 66 feet per second will change the angle of attack at 200 kts by as much as 11°, which will either lift you very quickly or cause a stall).

Here is a chart expressing angles of bank against load factors:

Angle	Factor
0°	1
15°	1.04
45°	1.41
60°	2
75°	4

The point to watch is that the load factor may increase so much that the maximum weight of the aircraft is exceeded, and the wings won't be able to do their job properly.

Propellers

These are just aerofoils with a twist in them (*washout*) to spread the lift evenly over the whole length, as the tips run faster than the center and need less angle of attack (the word *pitch* is sometimes used loosely to describe this). The basic propeller is averaged to cope with many flight conditions, so is not perfect for them all, particularly the takeoff. The real problem is that you have to make the engine run faster for more performance from the prop, and engines work best within a certain speed range.

Officially, a propeller's function is to convert the crankshaft's rotary movement into *thrust*, by moving a large column of air backwards, to propel the aircraft forward. As a propeller is an aerofoil, the thrust it creates is the same as the lift from a wing – it's just used differently.

A *tractor* (at the nose) will propel fast, turbulent air over the lifting surfaces, whereas a *pusher* (somewhere behind the fuselage) provides better high speed performance because it doesn't produce so much drag. On the other hand, the tractor bites into clean air, while a pusher spins in air that is already disturbed.

A rotating propeller creates various forces which may be allowed for in the design stages, including *gyroscopic precession* (see *Instruments*), where lifting the tail tends to make the nose yaw to the left. *Torque* results from the airframe going the opposite way to the direction of rotation:

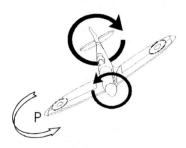

The effect is to produce a roll, which is countered by washout on the upgoing wing.

The blade going down pulls more at high angles of attack, resulting in *asymmetric thrust* (also known as the *P factor*). Where the propellers rotate the same way on a multi-engined aircraft (as with the PA 23), the failure of one engine may cause more problems than the other because of this – in the case of the Aztec, the downgoing blade is on the right side, since the blades rotate clockwise, so there is more of a turning moment if the left engine fails, as the thrust from the downgoing blade of the right propeller is further away from the longitudinal axis.

The left engine in this case is called the *critical engine*, because its loss creates the most adverse conditions. Later aircraft (such as the Navajo Chieftain) have contra-rotating propellers to deal with this (that is, the right engine rotates to the left). Watch for the designation of parts for particular engines with L or R so they go in the right place. The other thing to note is that the slipstream going back over the flight controls will also be asymmetric, and parts of

some controls will be more responsive than others.

Slipstream results from rotating air going round and round the fuselage until it eventually hits the tail fin, forcing it one way or the other (thus causing yaw), depending on which way round the propeller is going:

It can be reduced by offsetting the fin, as most of an aeroplane's life is spent in the cruise. Outside of that, simply use rudder.

Remember that, although torque and slipstream produce roll and yaw, they ultimately produce yaw and roll as secondary effects.

Propeller efficiency is the ratio of thrust to brake horsepower, the difference between what is available from the engine to what is actually used.

Geometric pitch is how far a propeller should move forwards in one rotation – the *effective pitch* is how far it actually moves. The difference between the two is *propeller slip*, and is a measure of the efficiency or otherwise of the process.

The *pitch angle* is that between the blade's chord line and its plane of rotation. The *helix angle* is between the resultant airflow and the plane of rotation. The surfaces of a propeller are the *thrust face* on the rear, and *pressure face* on the front.

Constant Speed Propeller

Otherwise known as a *Variable Pitch Propeller*, this performs pretty much the same function as the gearbox does in a car, in that it "maintains engine RPM over varying conditions of road", or flight, in this case. The gearbox (or constant speed prop) is there because engines work best within a certain range of RPM – going too fast or too slow is not good for them. In other words, a constant speed propeller can have its pitch adjusted for varying conditions, to maintain a constant angle of attack. Most are hydraulically operated with a *centrifugal governor* operating a control valve that lets oil in to make the pitch coarser or releases it for fine pitch (best for takeoff). Coarse pitch is used for the cruise because the propeller blades move a longer distance per rotation due to the higher angle of attack.

The *centrifugal twisting moment* is a component of centrifugal force that tends turn the blades into fine pitch. The *aerodynamic twisting moment* is the opposite, tending to coarse pitch (the blade's CP is in front of the pitch change axis). They both act in the same direction in a dive.

If the engine fails, and feathering is not available, select fully coarse.

Counterweights

These can be used to force the blades into increasing their angle of attack. The angle is reduced by oil pressure, so when oil pressure is lost, the blade angle will increase, although there may be an automatic feathering system.

With counterweights, therefore, loss of oil pressure feathers the blades. Otherwise, they go into full fine pitch (exam question).

Now you have more choice – you can operate at high RPM and low manifold pressure, and *vice versa*. Using lower RPM in the cruise (with higher manifold pressure) helps fuel consumption, since the engine is going round fewer times per minute, and the losses due to friction are less than at high RPM. However, using too much MP against low RPM will damage the engine (check the flight manual) and risk detonation.

The Helicopter

This is just a flying machine that has its wings going round instead of remaining still, cynically referred to by some as 50,000 rivets in loose formation – this means that the lift-producing surfaces (i.e. rotors) are separate from the body.

Another difference is that an aeroplane engine is directly related to the forward movement of the aircraft, whereas the engine on a helicopter isn't – its function is to drive the rotor system which is really what makes the machine move through the air.

In fact, the rotors provide lift, thrust and directional control in one go – all three being separate functions on an aeroplane. In some ways, this is beneficial, since loss of an engine on a twin does not need the strength of a gorilla to keep it straight, but it does make flying more demanding.

A helicopter's fuselage will have the tailboom and main rotors attached to it. The most common landing gear is skids:

But you may find wheels, too:

As you probably already know, a helicopter can move or turn in any direction, including up and down. The flight controls are the *cyclic* and *collective* controls, the *throttle* and the *tail rotor pedals*, which all have much the same effect as they do in aeroplanes (once out of the hover), except for the collective, which isn't used in them at all. As mentioned before, the main rotors provide lift, thrust and directional control and, because it is impractical to change the speed of the blades, or their shape, the pitch is altered instead when varying the thrust. Blade speed is constant, and the minimum and maximum speeds are close together, because of engine limitations on piston machines, and transmission limits for turbines (the minimum limits are for the coning angle).

The collective, to the left of the pilot's seat, is called that because it changes the pitch of all the rotor blades at the same time, that is,

collectively, thus changing their angles of attack all at once. Unfortunately, this also increases drag, which will tend to decrease the engine and rotor RPM, so some throttle needs to be applied when the collective is moved to keep them up (in fact, the throttle in a piston machine should be applied *just before*, so the engine doesn't lag behind, called *leading with throttle*). There is some sort of automatic linkage between the collective and throttle on most machines, but with pistons, this is rudimentary at best, and may not exist at all, as on the Hiller 12E.

Traditionally, the throttle is mounted on the end of the collective lever. Its function is to *regulate engine RPM*, and it is moved by the left hand outwards (away from the thumb) to increase power, and the other way to reduce it. Where (turbine) engine RPM is maintained by a *Fuel Control Unit* (FCU) or FADEC (*Full Authority Digital Electronic Control* – see below), it isn't moved at all, except in some emergencies where it can be used to control the direction of the fuselage. Because they are usually left in one position, turbine helicopters may also have the throttles mounted in the roof or on the floor which, of course, restricts their use when problems occur.

The cyclic control is the equivalent of the central column in a plane, and only changes the pitch of one blade at a time, to raise the rotor disk (or, rather, the *tip path plane*) at that point and tilt it in the direction you want to go. In other words, it changes the *direction* of the lifting force, and not its *magnitude*, except in the one place required to lift the blade.

Like the rudder in an aeroplane, the tail rotor pedals are *not* used to turn the ship (except in the hover), but to stop it turning the wrong way when you are turning, or to provide fine tuning for trimming purposes. In straight and level flight, or above about 60 knots, they can, to all intents and purposes, be ignored, as the tail boom does all the work.

The normal tail rotor, as found on JetRangers, requires a large number of components and sits in the dirty airflow from the main rotors – it therefore lives a stressful life. The *fenestron*, as used by Eurocopter, is a different solution, consisting of a series of very small blades enclosed in a shroud:

The blades are not equally spaced, to help with noise, and the shroud prevents tip losses, for more efficiency. Because the blades are rigid, they are less susceptible to vibration. They can also work closer to the stall, and their service life is longer because they are not so stressed – this also applies to the rest of the transmission.

The point about flying controls is that they should always be moved *smoothly*. Good helicopter flying is essentially downwash management, which has some lag built in. If you jerk the controls, you will get all the drag without the lift when the blades get into position before the airflow has a chance to catch up.

Control Orbit

The movements of the collective and cyclic controls are transmitted to the main rotors through a *swashplate* or *spider* system on the main rotor drive shaft, and their plane of rotation is called the *control orbit*.

With the swashplate, two circular plates are on top of each other, directly connected. Movement of the bottom plate is reflected directly in the upper one. The bottom plate is also stationary, except for sliding up and down the drive shaft, and connected to the cockpit controls. The top one rotates with the main blades and transmits the movements of the bottom plate to them.

The spider system uses an operating arm in a sleeve inside the main drive shaft with a ball at the end of it, that moves around a socket rotating with the blades, that itself is inside a stationary socket connected to the flying controls. Movement of the lower socket shows in the upper one, causing the operating arm to tilt and alter the pitch of the blades. Both sockets can move up and down inside the shaft so that collective movements are transmitted.

The *inverted spider* is as above, but upside down, on older machines.

With respect to a helicopter, lift and thrust together (or the sum of the lift of all blades) are often referred to as *Total Rotor Thrust*.

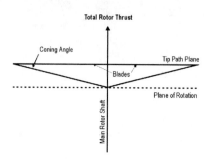

Lift is a vertical component, and thrust is horizontal:

The *Tip Path Plane* is the path described by the tips of the rotor blades, viewed horizontally. The

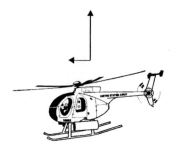

Coning Angle is the angle between the blade and the Tip Path Plane. The *Rotor Disc* is the area formed by the rotation of the blades, inside the blade tips. As the blades start *coning*, it will change its size slightly (the word *rotor* includes the blades, the hub and the shaft). The smaller the disk gets, the less area there is to generate lift – a situation that would arise if the RPM were too low, causing the coning angle to increase and centrifugal force to decrease. *Coning* is the resultant of centrifugal and lift forces – as the blades produce lift, the *coning angle* increases, but it decreases as RPM increases - the blades need centrifugal force for

stiffness, and their ability to support the machine in flight. It is measured in tons, against pounds for lift, thrust weight or drag, which will give you an idea of its importance.

Main rotor blades obey the same rules as for any aerofoil, with the exception of special shaping on modern machines to suit peculiar requirements (such as the Bell 407). Otherwise, they are symmetrical in cross section, to restrict movement of the Centre of Pressure.

The *pitch angle* is that between the blade's chord line and the spin axis of the main blades, and the *plane of rotation*, which is parallel to it (usually above). The plane of rotation contains the centres of mass of the rotating blades (the *axis* of rotation is a vertical element, which the blades rotate around – it is at right angles to the plane of rotation, and not necessarily in line with the rotor mast. The difference between them is the *flapping angle*). The pitch angle is varied with the collective and cyclic controls (see *Airframes, Engines & Systems*, next), and is *not* the same as the *angle of attack*, between the chord and the relative airflow.

Rotor Profile Drag comes from rotor blades at zero pitch, occurring purely because the blades are rotating. Air flowing through the disc at positive angles of attack suffers from *induced drag*, which is highest in the hover. The downwards motion of air through the blades is the *induced flow*.

As the helicopter moves, the blades will move above and below the plane of rotation, in a process called *flapping*. A *flapping hinge* allows this to happen, to cope with different angles of attack around the disc, and equalise the lift around it. They are needed when you have more than two blades, which would use a *teetering head*, and work like a seesaw for the same effect, where the two blades will flap as a unit (see below).

When the helicopter moves forward, the blade going forward will develop more lift because of the added speed (from the helicopter's forward movement and that of the blade), so it will fly higher. As it does so, the angle of attack reduces because of the change in relative airflow.

On the other side, the blade going backwards will generate a lot less lift because of its reduced speed, in some areas producing a reverse effect, which will cause the blade tip to stall if it gets large enough - on a Bell 206 at 100 kts, the non-lift producing area of the retreating blade is about 25%. This will make it fly down to increase its angle of attack, to create more lift (needing more forward cyclic to compensate).

Disymmetry of lift, therefore, is the difference in lift between the advancing and retreating blades, compensated for by flapping, which, unfortunately, causes the centre of mass of the blades to move, making them speed up or slow down relative to each other. Limited movement horizontally is provided with *dragging hinges* - *dragging* is the movement of a rotor blade forward or backward in its mounting. However, such hinges are only found in articulated heads (when a blade is ahead of its normal position, it is *leading*, and when behind, it is *lagging*). Semi-articulated heads (as with the AStar) may have a flexible coupling that allows fore- and-aft movement, but with no

flapping hinges – instead, the blades flex when compensating for lift. The pitch angle of the blades is changed by *feathering*, i.e. allowing them to rotate around their axes.

The speed at which the retreating blade tip stalls depends on the total pitch of the blade, that is, whatever is set by the combination of collective and cyclic. The cyclic input will increase with speed, and the outer part of the blade will stall first, the maximum effect being felt just aft of the trailing edge. In the cabin, you will detect a rolling tendency (usually towards the advancing blade) and a rearward tilt, together with stick and aircraft vibration and reverse cyclic behaviour.

Thus, the helicopter stalls as a function of going too fast, rather than too slowly, as with an aeroplane – the retreating blade flapping down to increase its lift gets a very high angle of attack, which announces itself with a lot of vibration. Try to avoid the problem if possible, by watching your airspeed and keeping away from V_{NE} in gusty conditions. Recover by lowering the collective.

The advancing blade can stall, too, but from *compressibility* and high speed buffeting when approaching the speed of sound, which will limit forward speed.

The rotor disc behaves like a gyroscope, and is subject to *precession*, meaning that an input doesn't have an effect until 90° later in the direction of rotation (see *Instruments* for more on this). Thus, if you pushed the cyclic forward, and the controls were not corrected, you would actually move left or right, according to which way round the blades were going. To cater for this, control inputs are applied in advance of the blades' movement. Their delayed response is *phase lag.*

The effects of this can be seen when increasing the collective in forward flight (say when taking off)– there will be a roll towards the advancing blade because the front portion of the disc is always more efficient than the rear, due to *Transverse Flow*, which is a fore and aft disymmetry of lift. When you raise the collective, the front portion of the disc creates more lift, which actually takes effect over the retreating side, causing a roll towards the advancing blades (right, in a 206).

The reason why the disc produces more lift at the front is because there is more induced velocity at the rear, and less angle of attack, and less lift.

Tail Rotor Drift

In the hover, the tail rotor provides more of a force than is actually required to counteract the torque from the main rotor. In other words, it's doing more work because it is impractical to place antitorque thrust at the front of the machine. In the picture below, the blades rotating around point O at points A are counterbalanced with a double force BB, as you would get with a typical tail rotor. If you cancel out one each of A and B, you are left with a side loading that causes movement:

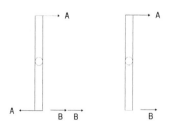

There is another way of looking at it, though. If you had contra-rotating main blades, the body would stay still, because the counteracting forces are in line with each other. The tail rotor, however is out on the end of the tailboom, and therefore has a moment arm, and enough leverage to cause movement.

The correction can be done simply by holding the cyclic slightly offset from its central position. Other ways include offsetting the mast or the engine, rigging the controls, or causing the disc to tilt when the collective is raised. None, however, eliminate it completely.

Tail rotor drift is why helicopters go one way or the other (depending on which way the blades go round) when the engine fails in the hover.

Disc Loading is calculated in lbs/square foot and is obtained by dividing thrust by the disk area. It doesn't change by adding more blades, or widening the existing ones, but blade loading is lowered.

Tail Rotor Roll

If the tail rotor is below the level of the main rotor, the drift mentioned above will cause a couple with the tail rotor thrust going the other way, causing one or other of the skids to be lower in the hover, depending on the blade rotation (it's the left one with *North American* rotation, that is, anticlockwise as viewed from the top). It is therefore totally normal for one skid to be lower than the other, unless you've left the refuelling hose in (actually, this characteristic is quite useful when landing on sloping ground, as long as the slope goes with the skids).

To combat this, you could raise the tail rotor on a boom or lower the rotor head, as is done with the Brantly, but the C of G position could screw that up anyway.

Tail Rotor Failure

When the tail rotor fails, it will be in varying degrees of positive, neutral or negative pitch, depending on what you were doing at the time, so if you can remember what it was, you will have an idea of the state of the pedals. Unless it's a drive failure, or you lose some of the components, the chances are that you won't discover the problem until you change your power setting, as it's very unlikely you'll be flying along in the cruise, for instance, and find a pedal forcing itself completely over to one side, as simulated by instructors on test flights, unless you have a motoring servo or similar, in which case your problem is hydraulics and not the tail rotor, although the effect might be the same. More typically, you will be in a descent, climb, cruise or hover, with the pedals where they should be and won't move when you want to do something else. When descending, for example, in the AS350, you will have more left pedal (more right in the Bell 206), both of which will aid the natural movement of the fuselage against the main rotors. The pedals would be in a neutral position if you were flying at medium to high speeds, and the power pedal would be forward in high-power situations, like hovering. In any case, the spread between the pedals is not likely to be more than a couple of inches either

way, certainly in a 206 – try an autorotation properly trimmed out to see what I mean. You will notice the same in the hover. My point is that the situation may not be as bad as frequently painted.

In fact, landing with a power pedal jammed forward is relatively easy, since the tail rotor is already in a position to accept high power settings (try also using a little left forward cyclic in a 206, and pivoting round the left forward skid), so you may be able to come in very slowly and even hover. If the pedals jam the other way (right in a 206), look for more speed because there will not be enough antitorque thrust available.

A drive failure, on the other hand, or loss of a component, will cause an uncontrollable yaw, and maybe an engine overspeed, so the immediate reaction should be to enter autorotation, keeping up forward speed to maintain some directional control (which is difficult in the hover, so try to get one skid on the ground at least), if you have time. If you lose a component, the C of G may shift as well, although an aft one in general has been found to help with this situation. Pilots who have been there report that there is a significant increase in noise with a drive shaft failure, and that the centrifugal force in the spin is quite severe. Anyhow, an autorotation is certainly part of the game plan, and as speed is reduced towards touchdown, you will yaw progressively with less control available in proportion, so it may be worth trying to strike the ground with the tailwheel or skid first (if you've got one), which will help you to keep straight—according to the

JetRanger flight manual, you should touchdown with the throttle fully closed, as you would if the failure occurs in the hover, to stop further yaw when pitch is pulled to cushion the landing.

However, in some circumstances, such as the cruise, sudden movements like this may not be the best solution. If you can reduce the throttle and increase the collective, this would reduce the effect of the tail rotor at the same time as keeping the lift from the main rotors, as does beeping down to the bottom of the governor range (difficult in most AS350s or Gazelles, where the throttle is not on the collective). The tail rotor is there to counteract torque, so if you give it less work to do, you will be more successful.

Otherwise, you might find a power and speed combination that will maintain height until you find a suitable landing area, then you've got as much time as your fuel lasts to solve the problem. Don't forget that the cyclic can be useful for changing direction and enabling you to fly sideways to create drag from the tail boom and vertical stabiliser, for example. It's the sort of situation where it pays to be creative sometimes. After all, the aim is to walk away, not necessarily to preserve the machine. Two other things you can try if you finally make the hover—stirring the cyclic so as to dump lift, and pumping the collective to produce a similar effect. Both will serve to confuse the machine enough so it forgets which way to turn! With a jammed power pedal (left, in a 206), what also works is to crab in the way the machine wants to, come to a high hover

sideways and let the machine settle by itself. You will find very little input is required by you.

If you want to run-on for landing, get the wind and/or nose off to the retreating blade side, so the fuselage is crabbing, and control your (shallow) descent with a combination of throttle and collective, applying more of the latter as the throttle is closed just before touchdown so you run on straight. Note that some helicopters (such as twins, or the AStar) won't let you use the throttle as precisely as that. Not only that, you may well be so busy that worrying about minor details like the wind's exact quarter will be the last thing on your mind. For a running landing, on most machines, about 30% torque at 30 kts will put you in a good position for landing at 30 ft, and a little power at the last minute will put your nose nicely straight. For the non-power pedal, keeping straight involves either more speed or less power, and you have to accept more of a run-on.

In an AStar (or TwinStar), the recommendation in the book is to come in with some left sideslip (i.e. crabbing right). Slow down until the nose starts to move to the left, and you have your landing speed.

Loss of Tail Rotor Effectiveness
This is sometimes known as *tail rotor breakaway*, or a stall, which is not strictly correct, as thrust is still being produced – it's just not enough for the task in hand. It shows up as a *sudden, uncommanded right yaw* (with North American rotation), and has amongst its causes high density altitudes and power settings, low

airspeeds and altitudes, and vortex ring. Your helicopter will be more susceptible to it if the tail rotor is masked by a tail surface, like a vertical fin, and it can be especially triggered by tail and side winds (this is actually a significant reason for maintaining main rotor RPM – as the tail rotor runs at a fixed speed in relation to it, lower NR will reduce tail rotor effectiveness in proportion). Recovery in this case comes from a combination of full power pedal, forward cyclic and reduction in collective, or autorotation. Prevention lies in keeping into wind and always using the power pedal (left in a 206 or one with similar blade rotation). If you use the other one, not only will the fuel governor ensure that the aircraft will settle after a short time (using the power pedal by itself makes it climb), but a large bootful of the power pedal in a fast turn the other way will create a torque spike.

Rotor Systems

Three or more blades require a *fully articulated* rotor, which essentially allows all of them to move in their various planes independently. This, however, adds complexity and expense to the design.

A *semi rigid* rotor has the blades fixed with regard to feathering, but they can flap up and down because the whole head can teeter, like a seesaw.

A *rigid* rotor only allows feathering, but the blades are more flexible towards their ends, so they bend when absorbing the forces of flight, producing the same effect as flapping and dragging hinges, but removed from the root.

In flight

In the hover, other things being equal, the lift vector acts directly upwards:

When you tilt the disc, the lift vector is reduced, because some of it is diverted to the direction selected:

The *resultant* (i.e. the diagonal line drawn across the two vectors) is where the main force finally ends up.

The *tangential velocity* is the speed of the blades' rotation. It increases with distance away from the hub, until it finally becomes a tangent to the edge of the disc, hence the name. Combined with the *downwash velocity*, you end up with a resultant corresponding with the blade's actual speed and path, or the relative wind (although its name suggests otherwise, the downwash component moves upwards):

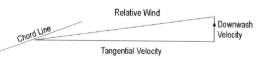

In autorotation, the function of downwash velocity is replaced by air going up through the rotors and creating a larger angle of attack.

The *flight velocity* is the reciprocal of the relative airflow, made up of downwash and tangential velocity and movement of the machine through the air.

Translation is the conversion from hover to forward movement, where the helicopter is supported by other means than its own power, that is, relative airflow. *Translational lift* is the extra thrust you get from forward movement, when the new airflow enters the disc. The helicopter flies better because you get more air through per unit of time, which has a lower induced velocity because it hasn't had a chance to speed up just before going through the rotor. As tip vortices are also being left behind, your lift vector becomes more vertical, for more thrust with less drag. The reason you have to lower the collective to maintain height at this point is because the angle of attack has increased against the new relative airflow. This also means less engine power is required. Of course, all the while rotor efficiency is increased with forward flight, at some point you need to increase power to overcome drag from the fuselage, which is increasing at a faster rate. This is

why you should not reduce power at the end of a climb until you have *both* the speed and height you want (if you reduced power at, say, 1500 feet and 60 knots, but you really wanted 100 knots, you wouldn't be able to accelerate beyond a certain point without applying more power than you would have used before).

All changes in velocity from cyclic movements are known as *transitions*.

Autorotations

Loss of RPM at the entry into autorotation is the most significant problem—a higher angle of attack from the new relative airflow as air rushes up through the rotors will cause enough drag to slow the rotors drastically, especially if your weight is high or the air density low, meaning that your blades will be at a higher pitch angle anyway, to meet the conditions of flight.

Reducing pitch to compensate will, of course, increase the rate of descent, at which point the inner 25% of each blade is stalled, and the outer 30% is providing a small drag force (in other words, it is being driven). The best lift/drag ratio in autorotation is at best endurance speed (check the manual, but most helicopters use about 45 kts). This is when the driving region of the disc is exactly centred. As you increase speed, it moves toward the retreating blade side of the disc until it touches the edge, which is your power-off V_{NE}. If it goes beyond the edge, the surface area of the driving region is reduced, resulting in rotor decay:

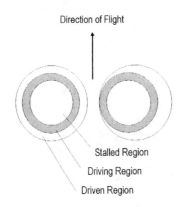

Direction of Flight

Stalled Region
Driving Region
Driven Region

When about 70 feet from the ground, (depending on whether you think you are descending or moving across the ground too fast), use rearward cyclic to slow down vertically and horizontally, in a manoeuvre called the *flare*. The amount is proportional to your speed and serves to increase the total lift reaction (which stops the sink) and shifts it to the rear (which stops forward movement). It also increases the rotor RPM. Continue the flare progressively (and sharply), to be at the correct speed for landing at 10 feet, applying collective as flare effect decreases to check the descent more positively, watching for drift (the "check" is the application of some collective to brake the descent – in the 206 it can be a positive movement; in the 407 and Astar, it can just be a pause). As the flare ends, and the kinetic energy of the rotors is used when the collective is raised, the airflow through the rotors is reversed, assisting you to level (the amount depending on the model), ready to cushion the landing as you apply collective pitch. This is where correct use of airspeed during the descent will have had the most beneficial effects—as the kinetic

energy stored in the blades is what slows you down, it follows that any you have to use to slow an unnecessarily fast rate of descent is not available for the final stages of touching down.

Ground Resonance

In flight, most parts of a helicopter vibrate at their own natural frequency. On the ground, they collect through the landing gear - if its natural vibration matches that of the main rotor, every time a blade rotates, the present vibrations receive another reflected pulse to increase their amplitude, and which can cause the aircraft to tip over and be destroyed. Peculiar to some helicopters, with fully articulated rotors, because they have dragging hinges (they are there to counteract vibration caused by movement of the blade's centre of mass), this is indicated by an uncontrollable lateral oscillation increasing rapidly in sympathy with rotor RPM. It could also be caused by blades not being in balance, unequal tyre pressures or finger trouble, but will only occur if the gear is in contact with the ground. It's best avoided by landing or taking off as cleanly as possible, but, if it does occur, you must either lift off or lower the collective and close the throttle.

Dynamic Rollover

This occurs when your helicopter has a tilted thrust vector with respect to the C of G, commonly encountered with some side drift when you have one skid or wheel on the ground acting as a pivot point, but you can also get a problem when your lateral C of G falls outside the width of the skids or wheels. Every object has a *static rollover angle*, to which it must be tilted for the C of G to be over the roll point, for most helicopters being 30-35°. As your lateral cyclic control at that point is a lot less effective than if you were hovering, because it is not rotating around the C of G, but the rollover point, you have less chance to get out of trouble, and the only effective control is through the collective (do not raise it). In other words, the lift from the rotor disc that should be vertical is inclined and converted into thrust, above the centre of gravity, so using the cyclic to level, and the collective to get you off the ground is *wrong!*

Dynamic rollover is worst with the right skid on the ground (counter clockwise main rotor) and with a crosswind from the left, with left pedal applied and with thrust about equal to the weight (i.e. hovering). The machine could roll upslope if you apply too much cyclic into the slope, or downslope if you apply too much collective, enough to make the upslope skid rise too much for the cyclic to control. The way to avoid it is to keep away from tail winds, and land and take off vertically.

Ground Effect

In the hover, downwash is stopped by any surface within about 1 rotor length. Because downwash velocity is reduced, so is the lift vector, which becomes more vertical (which itself increases thrust a little more and reduces drag), resulting in less induced drag and a reduction in the power needed to hover (a Bell 206 typically will need 15% less in the ground cushion). In addition, the accelerated air, having slowed down, increases the pressure underneath the rotor disc.

The effect is even more pronounced when you lower the collective to stop climbing, and will be more apparent closer to the ground. Factors that will reduce this are the surface you are hovering over (the harder and smoother the better, and the more level), and the wind, which will vary the direction of the downwash from under the blades. Any above about 10 kts, of course, will produce translational lift.

Recirculation

When hovering near the ground, some downwash comes back on itself and goes through the rotor disc twice, which reduces the lift whenever this happens because it does so at a higher speed and reduces the space available for the angle of attack in the resulting vector. Vortices are present at the rotor tips all the time, because they are caused by centrifugal force, but they are usually more than offset by ground effect.

If your ground effect is reduced for any reason (see above), the chances of recirculation increase, requiring more collective and cyclic to compensate. Where the downflow is actually prevented from escaping properly, as when hovering close to a building, or in a tight confined area, the effect will be to tilt you in a direction 90° from where recirculation was introduced, or even pull you down if all sides are affected (as when landing in a courtyard). Thus, if you are hovering a 206 next to a building in front of you, the recirculation occurs at the front, but the disc will tilt to the left and make the left skid hover lower than usual which, if it catches you unawares, might cause dynamic rollover (see above). If you are closer than a third of the disc diameter, the advancing blade is also affected, in the above example, pulling *towards* the building.

Be particularly careful within 1 rotor diameter of another helicopter.

Vortex Ring

This occurs when you encounter your own downwash, and you don't need a high rate of descent to do it (in essence, the vortices that should trail after you in the cruise remain around the machine at low speeds and interfere with your lift). The symptoms are random vibration, buffeting, pitching, yawing, rolling, an accelerated rate of decent and momentary loss of cyclic control. It is caused in a similar way to recirculation, since the airflow caused by the descent increases the blade vortices, which reduces the angle of attack as they share the same space. With a rate of descent matching the speed of the downwash, there is no angle of attack, and therefore no lift (the root area will be stalled). If you like, imagine the outer part of the main rotors encased in a large doughnut of recirculating air:

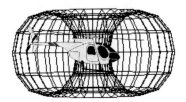

You are most likely to encounter it during a low or zero forward speed descent at a medium rate (500-1500

fpm) and a high power setting, typically found in a steep approach, where the column of air remains underneath your helicopter. To get out of it, reduce power, enter forward flight or autorotation, but bear in mind that you will lose a lot of height anyway. Better still, keep out of it with positive forward speed or by descending more gently, so the doughnut is below the machine.

Power

The types of power required are to overcome the various sorts of drag, namely *parasite power, rotor profile power* and *induced power*, which accounts for about 60% of the power needed to hover (rotor profile power takes up the other 40%).

Here is a graph that shows the relationship of power required against various forward speeds:

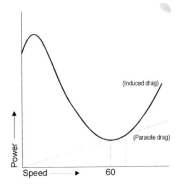

You can see that a little power is needed at first to prevent the machine sinking as the lift vector is tilted and reduced, and you transition into forward flight out of the ground cushion, then reduces drastically until the effects of parasite drag come into force and require much more power for forward speed (it increases as the square of speed).

The lowest point of the curve is the speed at which the least power is needed, and is therefore the best for *endurance*. The *maximum range* speed occurs slightly later, when the curve starts to rise more sharply (where the projected line from the origin cuts the curve a second time). The *best rate of climb speed* gives the maximum altitude in the shortest range, and because the maximum power is also available then, is the same as the endurance speed. In other words, the best ROC is obtained when there is the greatest difference between the power required for level flight and that available from the engines.

Blade Sailing

High winds and gusts will cause the main rotor blades of helicopters to flap up and down and be both a danger to people near them and the helicopter itself, as the blade stops could be damaged, or a particularly flexible blade could hit the tail boom. At certain critical speeds (50-100 RPM), blades will pass in and out of the stall. Holding the cyclic in the direction of the wind will keep the pitch of the advancing blade to a minimum and stop it lifting in the first place.

Other ways of minimising the effect include parking the helicopter away from the downwind side of obstructions or the downwash or slipstream of other machines, keeping the collective down, or accelerating and decelerating the blades as quickly as possible. In addition, point the nose out of wind, so that the lowest deflection is away from the tail boom:

Having the wind from the rear helps you keep an eye on the low blade at the front, but this means landing downwind in the first place.

Do not use collective pitch to slow the blades down – droop stops depend on friction for proper operation, and all you will be doing is lightening the load where it ought not to be.

Some Questions

1. What factors affect the amount of lift produced by an aerofoil?

2. What are the 3 axes that an aircraft moves around, and associated stability?

3. Why does an increase in all-up-weight lower speed in a helicopter?

4. What is the boundary layer?

Some Answers

1. The angle of attack, air density, velocity of airflow, surface area.

2. Longitudinal (roll) – lateral stability, lateral (pitch), longitudinal stability, yawing axis – directional stability.

3. More collective is required and the retreating blade will stall earlier.

4. The layer of retarded air immediately in contact with the aircraft skin.

Engines & Systems

Engines, turbine or piston, all work on more or less the same principle – a quantity of air is sucked in and mixed with fuel, compressed, set on fire and slung out of the back (suck, push, bang, blow in other words or, more technically, *induction, compression, power* and *exhaust*). The difference is that the power comes from the ignition stage in the piston, and the exhaust stage in the turbine, which is always ignited, whereas the piston only does so when the spark plugs operate. The jet is also a whole lot lighter, and spins a lot faster. In short, engines convert heat energy into mechanical energy, and not very efficiently at that (if they were, exhausts would be cold – the thermal efficiency of a piston engine is only 30%, although it does increase with altitude). The mechanical energy may be used to drive electrical, hydraulic and pneumatic systems as well, which is why engines are also called *powerplants*.

For short range aeroplanes, the propeller is the most economical method of propulsion, which can be driven by a piston or turbine engine (piston-driven ones are generally found on aircraft below 5700 kg). Because of inefficiencies in design, there is a loss of energy in the process, and the *thrust horse power* is about 80% of the *shaft* or *brake horse power* actually coming out of the engine. Sometimes, engine output is too great for the transmission, so the manufacturer will *derate* it to make sure it doesn't damage anything.

Reciprocating Engines

A typical piston engine consists of a series of identical *cylinders* which can be arranged in many ways, according to what the engine is going to be used for. The Beaver, for example had a Pratt & Whitney R985, which was a *radial* engine that had the cylinders in a circle, attached to the *crankshaft* in the centre (with the propeller bolted to it). In this case, the cylinders stayed still, and the crankshaft moved, but earlier engines made the crankshaft stay still while the cylinders moved. As you can imagine, maintenance was difficult, to say the least.

Most modern light aircraft, including helicopters, have their cylinders opposite each other (*horizontally* or *vertically* opposed), to cancel some forces out, but many have them in line and even in a V formation. Some engines are even upside down, as found with the Gypsy Major in the Chipmunk (since the pistons in this case are at the bottom, there is a danger of oil leaking into the combustion chambers, which is why the engines are turned through one complete cycle before being started, to make sure the insides are free to move. You will break the starter if the cylinders are blocked with oil).

One big difference the average car driver will find is that an aero engine has two spark plugs per cylinder, which are powered by independent *magnetos* (see below). When doing power checks before takeoff, they are checked against each other for power and whether they actually are independent. Another is that car engines are a lot smoother (there are some diesel engines that won't upset a coin standing on its edge). As cylinders in an engine are all the same, we will look at just one in a moment to see how they work (see *The 4 Stroke Cycle*, below).

The *cylinder* is just that, but it is closed at one end to provide an airtight seal (there are *valves* in the *cylinder head* to allow the mix of air and fuel in, and exhaust gases out, but only at specified times). Inside the cylinder is a *piston*, which slides up and down to provide an action like a pump, since it pulls air and fuel in, and pushes the exhaust out.

As the piston is meant to be gastight, and no fit is perfect, the piston will have two or three rings round it (the *scraper* is for cleaning) to mate against the cylinder wall and stop movement of anything from one side of it to the other because, on the one hand, the engine will not produce full power if the burnt gases leak out and, on the other, oil will get through to the head from the lubrication system, mix with the fuel and air and cause a lot of bluish grey smoke (if you are getting mysterious oil leaks from your car, and everything appears to be done up underneath, check your piston rings, as they may be allowing pressurised gases through to the sump to force the oil out).

The piston is attached to the *crankshaft* with the *connecting rod* (or *conrod*, for short), with a *big* and a *small* end, for the crankshaft and piston, respectively (if either end goes, the engine will suddenly start clattering loudly):

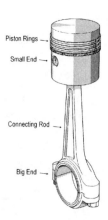

Piston Rings →

Small End →

Connecting Rod →

Big End →

The crankshaft in an in-line engine (that is, one that is not a radial, including V8s and horizontally opposed) is not straight– it is offset for each piston connected to it, one after the other, so the up and down (reciprocating) movement of the piston is translated into rotary motion, to drive a propeller or main rotor gearbox:

The length of the piston's travel inside the cylinder is known as the *stroke*, and the peaks of its travel up and down are known as *top-* and *bottom* dead centre. The former is an important factor in the timing of the spark, discussed below.

When the piston is at the top of its stroke, there is a small space between it and the cylinder head, into which the fuel/air mixture is pulled and compressed (it's called the *combustion chamber*). In being compressed, the mixture gets warmer, making it more disposed to ignite when the spark plug fires:

Unfortunately, some mixtures get so warm they can ignite without the spark (diesels work this way), which will not only increase the operating

temperature unnecessarily, but cause real harm to the engine from shock waves, caused when the piston gets the effects of the power stroke when it doesn't expect it (pound for pound, fuel is more explosive than dynamite). This *pre-ignition* also comes from hotspots in the cylinder, such as the end of the spark plug, or lead oxybromide deposits that result from leaning, which are likely to be glowing bright red. This may happen if cooling is inadequate, as when climbing too steeply, and is why an engine sometimes doesn't stop.

Once the mixture has ignited this way, the mixture explodes, rather than expands smoothly – in other words, *detonation* is caused. It's otherwise known as *pinking*, because it sounds like that. You will hear it in your car if you make it work too hard (try going uphill in high gear). Detonation can cause the temperatures inside the cylinder head to rise to the melting point of the components inside it, with the piston usually going first. The hot gases will leak past the piston rings, pressurise the crankcase and blow the oil out. Net result: seized engine and holes in the pistons. Time to open the wallet.

Because of these problems, engines use fuel with an *anti-knock additive*, which used to be lead, to ensure fuel ignites smoothly, and doesn't explode, and to stop it igniting before it's meant to (in the days before carburettors, fuel was much more volatile, and could be ignited ten feet away). Lead, of course, is no longer politically correct so, in cars, the timing of engines is adjusted to produce the same effect with unleaded fuel.

The aviation industry still uses it, though. The "LL" in 100LL stands for *low lead*, but there is still about four times more than is needed. As well as the lead (in the form of TEL—*Tetra-Ethyl Lead*), a scavenging agent (*Ethylene DiBromide*, or EDB) is also added to ensure that the lead is vapourised as far as possible, ready to be expelled from the cylinder with other gases. Unfortunately, this is not 100% successful, but the results are best at high temperatures and worst at low ones - the unwanted extras result in fouling of spark plugs, heavy deposits in the combustion chamber, erosion of valve seats and stems, sticking valves and piston rings and general accumulation of sludge and restriction of flow through fine oil passages, so it makes you wonder which is worse.

Anyhow, the *octane rating* reflects the ability of fuel to expand evenly. Aviation fuel is coloured this way:

Colour	Fuel
Red	80/87
Blue	100LL
Green	100/130

The smaller the combustion chamber, the higher the *compression ratio* of the engine, which is actually the difference between the chamber and the stroke of the piston or, in other words, the capacity of the chamber with the piston at each end of its stroke. Pistons generally have a concave surface at the top.

Aero engines are usually cooled by air, using the flow caused by forward movement, but some have complete cooling systems if the engine is packed into a tight space that air

finds it difficult to move through (the Spitfire, for example, used large amounts of glycol).

The fins outside a cylinder head are there to increase the total surface available to the cooling airstream.

Timing
The cylinder head contains valves which must be opened and closed at precise times to allow the fuel/air mixture in and exhaust gases out (*fuel* or *inlet valves* and *exhaust valves*):

Valves are hollow, or partly filled with sodium to encourage heat transfer.

As it turns, the crankshaft will turn a smaller version of itself (called a *camshaft*, which rotates at half the speed), linked directly to the *valve rockers* at the top of the cylinder with a long metal rod. The bottom end of the rod is enclosed in a *tappet* (to save wear), and the top end hits the valve rocker directly, pushing the valve open. As the engine gets hotter, these rods expand, so there is a little clearance to allow for this, called the *valve rocker clearance* (valve rockers are *not* tappets).

There are reasons why it's a good idea to open valves at a different time than top dead centre – one is that it helps with unleaded fuel, as mentioned above. Another is that an engine is complicated, with a lot happening in a short time, and some anticipation here and there doesn't go amiss. Opening early is called *valve lead*, and being late is called *valve lag*. When open at the same time, you get *valve overlap*.

Ignition is automatically *advanced* as RPM increases, and *retarded* when a light aircraft starts up (the spark is intensified as well). This is done with an *impulse starter*, which uses a coiled spring, or a *low tension booster* and *retard breaker*.

The Carburettor

This is a device that mixes fuel and air in the correct proportions, vapourises it and delivers it to the cylinders via the *inlet manifold* (that is, an inlet that serves *manifold*, or *many*, cylinders). It makes use of the *venturi principle*, already discussed in *Principles of Flight*, which states that as the speed of air increases over a restriction, the corresponding pressure reduces.

The *fuel nozzle*, which is connected directly to the fuel system, is inside the low pressure area, so the fuel in the line is sucked out and vapourised as it is forced to expand (this also cools the area, so be careful with carburettor icing, which can form well in advance of any other type). In fact, if you could make one small enough, there's no reason why you couldn't use an air conditioning unit to achieve the same effect.

Just before the carburettor ends and the inlet manifold begins is a *butterfly valve*, which is controlled directly by the throttle. When the throttle is

closed, the butterfly valve is closed, and *vice versa* (although better vapourisation and atomization can be obtained if fuel is introduced through holes in the butterfly – check out the *Fish Carburettor*, which was invented to stop fuel wastage when a car was thrown around).

Even when the butterfly is fully open, though, there is still resistance to the flow of fuel from its sideways presentation. New car engines have eliminated it altogether by making the throttle increase the inlet valve opening time to get the same effect.

Anyhow, when the butterfly is closed, the engine still needs to be fed with fuel, so there is an *idle jet* that bypasses it to keep the engine idling. It also helps the venturi, since the airflow at idle is quite small (the jet is actually a hole next to the butterfly, and it's sometimes called the *slow running jet*). Also, when you need power in a hurry, there is a small lag due to inertia between the time you open the throttle and the time the engine starts to speed up, because the air supply responds more quickly than the fuel, which gives you a *weak cut* (a momentarily weak mixture), so a small squirt of fuel is delivered separately to compensate for this, from an *accelerator pump*. When starting an engine from cold, therefore, resist the temptation to pump the throttle, because all you will do is flood it with large drops of fuel. A better tactic, if you need the throttle open, is to do so *very slowly*, so the pump doesn't kick in.

Because aeroplanes go up, and because air gets less plentiful at height, there is a danger of the fuel/air mixture getting out of balance as you climb. A mixture that has too much fuel against air is *rich*, while one the other way round is *weak*. The *mixture control* is provided to adjust for this – for example, you would have it set fully rich for takeoff and landing. At height, the engine will not work at all if the ratio of fuel to air is not correct.

Leaning makes the engine run hotter and give you more power for less fuel; a 112 hp aircraft cruising at 4000 feet and 85 knots will burn 5 gallons an hour when rich, but only 4.5 when leaned, giving a range of 116 miles as opposed to 100—a saving, or an increase, of 16%.

Although you could lean off slightly in the climb for better economy, never take off with reduced power or too lean a mixture. It may save fuel, but petrol has a high latent heat content, and the excess inside a cylinder from a rich mixture has a cooling effect when it evaporates.

The "normal" mixture is about 15:1 of air to fuel by weight, but this is not critical over quite a wide range (some say 14:1 is correct).

The mixture control has a secondary function, which is to cut fuel from the engine on the ground when you want to stop it (you don't just switch the magnetos off). The *Idle Cut Off* (ICO) in the carburettor is joined to the mixture lever with a Bowden cable. When the lever is operated at the end of a flight, the engine is starved of fuel, and stops.

Carburettor Icing

This is actually one aspect of *induction system icing*. The other two are *fuel icing*, arising from

water suspended in fuel, and *impact ice*, which builds up on the airframe around the various intakes that serve the engine. Even on a warm day, if it's humid, carburettor icing is a danger, especially with small throttle openings where there's less area for the ice to block off in the first place (as when descending, etc.). Also, the temperature drop (between the OAT and that in the venturi) can be anywhere between 20-30°C, so icing (in an R22, anyway) can happen even when the OAT is as high as 21°C (70°F), or more. Tests have produced icing at descent power at temperatures above 30°C, with a relative humidity below 30%, in clear air. Because it is more volatile, and likely to contain more water, you can expect more fuel and carb icing with MOGAS than AVGAS.

It usually arises from the action of the venturi in the throat, just before the butterfly valve, which regulates the amount of fuel into the engine. You will remember the venturi's purpose is to accelerate airflow by restricting the size of the passageway, which has the effect of reducing the pressure and pulling the fuel in. Unfortunately, this process also reduces the temperature, as does the fuel vapourisation, hence the problem (the lower temperature means greater relative humidity, and closeness to the dewpoint, and the vapourisation takes its latent heat from the surroundings, making the situation worse). In fact, the vapourisation (and cooling) can carry on most of the way to the cylinders, causing the problem to persist, especially with the butterfly semi-closed, which produces another restriction and more of the same. Any water vapour under those conditions will *sublimate*, or turn directly to ice. Note also that warm air will produce more ice because it can hold more moisture.

With smaller engines, use *full settings* for every application— that is, carb heat either on or off, with no in-betweens - the greatest risk is at reduced power. *Out of Ground Effect* hover performance charts for helicopters usually assume the carb air is cold (the R22 requires carb heat below 18" MP). In fact, when heat is applied, an engine will typically lose around 9% of its rated power.

Rough running may increase as melted ice goes through the engine. Also, be careful you don't get an overboost or too much RPM when you reselect cold. Of course, aeroplanes have some advantage if the engine stops from carb icing, as the propeller keeps the engine turning, giving you a chance to do something about it. In a helicopter, due to the freewheel that allows autorotation, the practice of only selecting hot air when you actually get carb ice may not be such a good idea – usually, a gauge is used with a yellow arc on it, showing the danger range:

Use carb heat as necessary to keep out if it. The other peculiarity with regard to helicopters is that they tend to use power as required on takeoff, whereas aeroplanes use full throttle. This makes them more vulnerable, as the butterfly opening is smaller, and is particularly apparent on the first takeoff of the day, when the engine and induction system are still cold. If it is filtered, your carb heat may be used to preheat the induction system during the engine warm-up. With a fixed pitch propeller, the first indications will be a slight loss in RPM, followed by rough running, then more loss of RPM until the engine stops. With a constant speed propeller (and in a helicopter) keep an eye on the manifold pressure gauge and the EGT gauge, if you have one, which should show a decrease.

Fuel Injection
Most of the above problems with the carburettor are avoided with fuel injection, where the fuel is metred directly to the cylinders according to power requirements, automatically taking air density into account. Ice is not formed because there is no venturi to cause temperature drops

(there's no carburettor in the first place, as it is replaced with a *fuel control unit*). In fact, the only control you have (apart from the throttle) is *Idle Cut Off* (ICO). The biggest problem with fuel injected engines is blocked jets, from dirt in the fuel.

A lesser one, but still significant, is difficulty in starting, particularly on a hot day, where the feed pipes lie across the top of the engine and consequently get warm (like the PA 31), with the fuel inside them evaporating nicely and creating a *vapour lock*, so you need a short burst of fuel pressure from the pumps to prime the lines.

If the primary airway is blocked (maybe from ice, or a bird), there will be an *alternate air* switch in the cabin to change the source.

An electric fuel injection system will be backed up by a mechanical one. The benefits include being able to use it without the engine running, making it useful for priming.

Superchargers

A supercharger is a compressor run directly by the engine, situated between the carburettor and the inlet manifold. Its function is to *extend the service ceiling of the aircraft*, by compressing the fuel/air mixture to maintain sea level power at altitude, or to increase normal power lower down. This *forced induction* only really works in a particular temperature range. The essential point to remember is that the extra air is *sucked through* the carburettor and *blown into* the cylinder.

Turbochargers

A turbocharger, on the other hand, is powered by exhaust gases which are deflected from their normal course outwards by a *wastegate*. When the wastegate is *closed*, the engine is being turbocharged. The gases drive an impeller that also compresses air, but at the *intake* (before the carb, so the air is being *blown into* it). Also, the engine doesn't lose power by driving it. Since the exhaust system is involved, preflight checks should include security of the pipes, so carbon monoxide doesn't get into the cabin.

Automatic waste gate control is done with engine oil pressure acting against a spring, which opens the gate at low altitude. Oil closes it at high altitude, although, at low engine power it will be closed anyway, to conserve pressure. The *critical altitude* is the pressure altitude at which it is fully closed. Above it, power will fall in line with the manifold pressure.

The *waste gate actuator controller* is downstream of the actuator.

Turbocharger bootstrapping (or *hunting*) is an overreaction to rapid throttle movement, which leads to large pressure fluctuations and possible overboosting. A pressure relief valve in the induction manifold cures it. The turbocharger controller can also be damped.

Turbocharger intercoolers (between the compressor and throttle) are there to prevent detonation at high altitude.

The turbocharger is lubricated with engine oil.

Ignition

This is the whole mechanism that provides the spark at the critical moment, consisting of plugs, leads, magnetos, switches, etc., in duplicate (one magneto will serve one plug per cylinder, and the second the others). The duplication is actually for efficiency, as the magneto doesn't work that well at low RPM, but a side benefit is, obviously, safety (watch the RPM drop as you switch to single mags in the runup to see what I mean about efficiency).

The magneto is a device that has a transformer and all the circuitry to boost the low voltage primary current to one large enough to jump across a small gap at the plug electrodes (around 15,000 volts). A car has similar items, but not in one unit. The difference is that a magneto works as long as the engine is turning it – there is otherwise no need for external influence.

According to the *Electricity & Radio* chapter, this is done with the help of *magnetic induction*, or the relative movement of a magnetic field around a conductor, in which a current is induced. On the other hand, if a current flows in a conductor, an associated magnetic field varies with the current flow.

In a rotating magnet magneto, a *rotor* rotates in the gap between two ends of something like a horseshoe magnet, called an *armature*, the bulk of which is surrounded by two sets of coils, a *primary* and a *secondary* winding. As the rotor spins, the flux along the armature reverses as the lobes pass the ends (in the above case, you will get four flux reversals

and four sparks per revolution, so it runs at ¼ of the engine speed).

The *primary coil* oscillates in sympathy with the flux reversals, and a current is induced in the *secondary coil* wrapped around it (because the primary field is expanding and collapsing – the essential point is *movement*). The ratio between the coils is significant, which is how such a high voltage is generated. The primary has thicker wire and generates about 50 volts, but the secondary wire is hair-thin and produces over 15,000.

Inside the primary circuit is a set of *contact points* that act as a *circuit breaker* whose function is to make the flux reversals more abrupt, since they are not good enough to produce a clean signal by themselves. In short, the points are there to *make* and *break* the primary circuit.

There is a cam which moves a rocker arm to the open position as many times per revolution as there are lobes on the cam. There is a spring to return the rocker arm to its normal position. The separation of the points needs to take place in strict timing with the maximum primary current, so the field collapses at maximum intensity.

As the points separate, a spark is generated across them as the primary current ceases, due to a strong follow-on current in the primary (electricity has a momentum). A condenser is used in parallel with the points to minimise the spark, but the points still need to be looked at for pitting and damage because it can never be eliminated entirely.

The high-tension current generated is fed to the plugs by a *distributor*, which is essentially a rotor spinning inside a cap holding heavy cables going to the plugs (part of the same unit as a magneto, but separate in a car). The one in the middle is the *King Lead.* When fitted, they are not matched to the plugs in order, as one cylinder would receive the spark at an entirely wrong time and strain the crankshaft. Instead, the cables are arranged out of order, on a four-cylinder engine as 1342 or 1243 (you will often see these numbers moulded into the cylinder head as a reminder, in case you were wondering what they were for).

The switches in the cockpit ground magnetos to Earth through the primary circuit, because they cannot be switched off (they work as long as they spin), so *they must always be treated as live.* Ground connections can fail.

The 4-Stroke (Otto) Cycle

Now we are acquainted with them, here, in excruciating detail, is a description of the movements of all the parts in an engine.

It all starts with the piston at top dead centre, ready to move down and suck in a fuel/air mixture from the carburettor, through the fuel valve, which has just opened:

On top of the suction created by the piston's downward movement, atmospheric pressure helps to force the fuel and air in (when it is less, the supercharger or turbocharger helps). The valve closes as the piston reaches bottom dead centre, so the chamber is filled.

With both valves closed, the piston starts moving up again, compressing and heating the mix, as well as increasing the *density,* which helps the flame ignite quicker because the particles are closer:

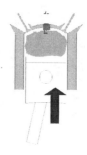

Just before TDC, the spark plugs (in the cylinder head, near the valves) ignite the mixture with sparks from a high-voltage electric current provided by the *magneto,* which is rotating in sympathy with the engine:

It is timed this way to give the fuel time to catch fire, and produce the optimum expansion at 10° after TDC, which is when it is actually required. The piston is forced downwards again, in a smooth movement, making the crankshaft rotate, and whatever is attached to it.

The *volumetric efficiency* is the measure of mass charge to the theoretical mass charge at ISA if the engine were stationary. In other words, the degree to which the cylinder is filled with new mixture at full throttle, as compared to an equivalent amount of atmosphere. It is rarely more than 80%, due to various leakages and losses, hence the need for supercharging, above.

The momentum of the engine (supplied by the *flywheel*) brings the piston up again, to force the exhaust gases out through the exhaust valve, which has already opened, just as the piston hit BDC:

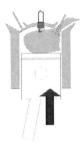

It closes as the piston gets to TDC, after the fuel valve has opened early, to allow some atmospheric pressure in that will help push the exhaust out, as the piston by this time is starting to slow down, ready to reverse direction.

There – that wasn't so bad, was it? Note that, although there were four cycles, the crankshaft only went round twice.

Turbines

As mentioned above, the same principles apply to jet engines as reciprocating ones, only they're applied in a different way. They also use cheaper fuel, as compression is not a factor in producing the power, although avgas can sometimes be mixed with jet fuel (see the Flight Manual), at the expense of reduced maintenance periods, as it doesn't the fuel pumps so well.

Turbine engines are discussed in this book only in relation to helicopters, because it is more likely that somebody with a Private Pilot's Licence is going to fly one.

Because so much air is used for cooling, humidity has less of an effect on jet engine performance.

The five basic parts of a jet engine are the *inlet*, the *compressor, combustor, turbine* and *nozzle* (the bit joining the compressor to the combustor is called the *diffuser*). They could be combined or doubled back on themselves in some engines, particularly those in helicopters, to save space, speaking of which, remember that the engine is not directly responsible for pushing the machine forward, as it might be in a jet aeroplane. Instead, it drives the main rotor gearbox, which drives the rotors – the disk formed by them is what flies and pulls the rest of the helicopter with it. In such circumstances, the engine could even be upside down, provided the gearing could cope with it. In fact, the PT6 has a "backwards" design, with the compressor at the rear.

Another name for a jet engine is a *gas producer* because, in a helicopter, the stream of hot gases is intercepted by a turbine, and used to drive a rotor gearbox.

About 2/3 of the energy produced is used simply to keep the engine running. Most of the rest is used by a power turbine for propulsion, leaving enough energy to ensure the gas falls out of the engine by itself, so you don't need extra components that will drain more energy.

The Inlet
Strictly speaking part of the airframe, this is where air enters the system. Its function is to convert ram-air pressure (from forward movement) into static pressure, ready for the compressor.

The air travelling through the inlet may well include other odds and ends, like sand (in the desert) dust, leaves, etc., especially in a helicopter, when you will be in the lower parts of the atmosphere anyway, and more prone to *foreign object damage*. Fine screens are used to combat this, but they do restrict the airflow and have an effect on your performance. Another device is a *particle separator*, which uses centrifugal force from inlet air to create small swirls that pick up small particles and drop them into a *sediment trap* (that is, rather like a vacuum cleaner). They work with snow as well.

The Compressor
This is a rotating mass of impellers or blades, designed to take vast quantities of air, compress it (and therefore heat it) for direction to the combustor (below), so it's an air pump, sometimes with the weight of air delivered determined by the engine RPM. That is, for any

specified RPM, the air volume will be a definite amount. The temperature rise across the compressor could easily be 555° (as on the Bell 407) and the compression ratio nearly 10:1 for a centrifugal compressor, and 25:1 for an axial (which means more thrust for the same frontal area).

The compressor can be *centrifugal*, or *axial*, or both. As its name implies, the centrifugal type uses impellers, as used with water pumps, to fling air outwards into channels leading to the combustion chamber. The axial compressor is essentially a series of wheels in line with each other, having fan blades around the outside of each one. The blades used to be attached separately, but now a complete wheel is created, with blades, out of one crystal. The air is forced back into stationary *stator blades (or stator vanes)*, to alter the characteristics of the flow – in fact, the pressure is gradually increased as it is forced into the smaller spaces created by further blades downstream. Each rotating wheel with its set of stationary blades is a *stage*, so several together (on the same shaft) would constitute a *multistage compressor* (the same thinking applies to turbines, below).

A *dual compressor*, on the other hand, would have stages in tandem, but on different shafts at different speeds, to produce higher compression ratios. The first in line would be the *low pressure compressor* (N_1), driven by the *low pressure turbine*, which would also be the slowest, via the *low pressure shaft*, which rotates inside the high pressure shaft, which performs the same function for the high pressure compressor (N_2) and turbine (in a helicopter, N_1 is also called Ng, and N_2 is also called Np). The N_2 shaft runs the opposite way to N_1, so the torques counteract and cancel each other out, relieving stress on the engine mounts.

The whole combination of shafts and compressor is known as a *spool*.

The *compression ratio* is the difference between the pressure of the air as it comes out of the compressor and the pressure at the *engine* inlet – it should always be higher than the back pressure from the turbine, or the airflow through the engine could go the wrong way.

Inlet Guide Vanes adjust air going into the compressor, which are closed when the engine is idling and fully open at about 70% engine RPM.

Some engines have a small valve that opens when the engine starts, to correct the airflow so that the compressor blades do not stall (a *compressor bleed*) - for maximum efficiency, and because engines have to react quickly, you need to operate as close to the stall as possible. At low RPM, the engine is naturally is not able to pump as much air, so you need to "unload" it during start and low power operations. A bleed air system makes it see less restrictions by staying open until a certain pressure ratio is obtained. Other engines may use such a bleed to prevent stalling when the throttle is opened too suddenly.

A *compressor stall* reduces efficiency, meaning less power. A *cold stall* only affects a few blades or a small area of the compressor, whereas a *hot stall* involves it all, and may mean severe damage caused by hot gases from the combustor when the airflow becomes reversed inside the engine. There may or may not be a loud noise to accompany this.

In fact, the compressor is an ideal place from which to tap small amounts of air (*compressor bleed air*) for other purposes, such as cooling, pressurising of oil systems or operating anti-ice systems. However, when doing this, the exhaust temperature will tend to rise slightly. For anti-ice systems, the bleed is taken from the back end of the compressor (that is, *compressor discharge air*) that has already been heated due to compression. It will typically flow through the compressor shell and hollow struts, and the inlet guide vanes.

The Combustor

The air is divided into two streams here, one for burning, and the other for cooling. Once the flame is lit by the *spark igniter* during engine start, it stays that way till the engine is shut down (you could say that the engine is on fire all the time, and it's only when the fire becomes uncontained that it becomes an emergency). The engine is spun initially by air from a high pressure bottle or APU (or even a V8 for the SR-71, or a starter/generator in helicopters) and the sparks ignited when the airflow is high enough to keep the temperatures down at the back end.

An *auto relight* system is designed to restart the engine should it flame out, providing a continuous spark from the *igniter* all the time it is switched on.

The Turbine

This is where the hot air flows through. As it spins, it also helps drive the compressor, as it is directly connected.

In a *free turbine* engine, used in helicopters and some turboprops, the exhaust goes through two turbine stages, e.g. a *compressor turbine*, and a *power turbine*. That is, there is no direct connection between the exhaust from the engine and the gearbox it drives, which is important for helicopters that need to autorotate when the engine stops. There is always a reduction gear system to reduce the high RPM coming from the power turbine (rotor blades, for example, only go round at about 300-400 RPM, whereas a jet engine will be more like 33,000). A helicopter powered by a free turbine may have a braking system to keep the blades and transmission stationary.

One of the most important instruments in your cockpit is the *Turbine*, or *Exhaust Temperature* gauge, which shows the heat coming out of the back end. It is particularly important during starting because, if the battery is too weak to spin the engine properly, there will be less airflow through it, and not as much cooling available, leading to a hot start and an expensive repair as the back end melts. During flight, on hot days, this temperature may well be the limiting factor in the amount of payload you can take, even if you have lots of torque left.

Jet Fuel

Jet A, standard for commercial and general aviation, is narrow-cut kerosene, usually with no additives apart from anti-icing chemicals. **Jet A1** has a different freezing point and possibly something for dissipating static, used for long haul flights where the temperature gets very low. **Jet B** is a wide-cut kerosene with naphtha in, so it is lighter and has a very low flash point (it's actually 2/3 diesel and 1/3 naphtha, but in emergency you can swap the naphtha for avgas to get pretty much the same thing). It contains static dissipators. *Try not to mix Jet A and Jet B* - the mixture can ignite through static in the right proportions, as Air Canada found when they lost a DC-8 on the ramp in the 70s. Static can come simply from the movement of fuel through the lines. Jet A weighs about 5% more per litre than Jet B, but it gives you a longer range, as turbines work on the weight of the fuel they burn, not the quantity. So, if you load the same amount of fuel, your machine will weigh more with Jet A, but if you fill the tanks, you will use fewer litres and less money.

JP4 is like Jet B but also has a corrosion inhibitor and anti-icing additives. It was the main military fuel but is being superseded by JP8, at least in the USA. **JP5** has a higher flashpoint than JP4, and was designed for US navy ships (similar to Jet A). **JP8** is like Jet A1, but has a full set of additives.

FCU

The *Fuel Control Unit* does more or less the same job as a carburettor on a piston engine, but it uses springs and bobweights to metre fuel according to demand.

FADEC

The initials stand for *Full Authority Digital Electronic Control*. It's just a computer that controls the fuel system, based on information from various sensors, such as exhaust temperature, engine RPM, control movement, etc. The end result is a more precise control of rotor speeds under varying flight conditions, particularly with reference to overspeeding. Other benefits include automatic starting, better care of the engine (so more time between overhauls) and reduction of pilot workload through automation. Being a computer, it is software-based, and one of the preflight checks is to ensure that the right software is loaded. Also, because it's a computer, it's able to monitor many parameters, which is why you might see more caution lights.

It will typically consist of two main items, the *Engine Control Unit* (ECU), on the airframe, with a processor inside (e.g. a 486 – powerful, huh?), and the *Hydro Mechanical Unit* (HMU) on the engine, which functions rather like the old-style FCU when the FADEC is disabled. There will also be sensors and relays for the transmission of information around the system. Many signals will be repeated to the relevant instruments.

Engine Instruments

Refer to the *Instruments* chapter.

Engine Handling

One of the biggest things to unlearn when transitioning from piston to turbine is to keep your finger on the starter button once things start happening (with a piston, you tend to take your finger off straight away

when the engine starts). You take your finger off when the engine becomes *self-sustaining*. Before then, it relies heavily on the battery or APU to keep it turning. It follows that, if the battery is weak to start with, the engine won't spin as fast, the airflow is reduced, the whole process becomes hotter and you could melt the back end with a *hot start*. You should always check the voltage available from the battery before starting a turbine engine. A *hung start* exists when the engine fails to accelerate to normal idle RPM. It just sits there, weakening the battery and leading to a hot start. You get a *wet start* when the engine doesn't light off at all (flooded).

Pulling full power just because it's there is not always a good idea. Limitations may be there for other reasons—for example, the transmission might not be able to take that much, which is why you can't go faster than 80 kts in a Jetranger when pulling more than 85% torque (actually, in this case, the transmission ends up in a strange attitude). Excessive use of power will therefore ruin your gearbox well before the engine (and will show up as metal particles in the oil). *Maximum Continuous Power* is the setting that may be used indefinitely, but any between that and maximum power (usually shown as a yellow arc on the instrument) will only be available for a set time limit.

While I'm not suggesting for a moment that you should, piston engines will accept their limits being slightly exceeded from time to time with no great harm being done. Having said that, the speed at which the average Lycoming engine disintegrates is about 3450 RPM, which doesn't leave you an awful lot of room when it runs normally (in a Bell 47, anyway) at 3300! Turbines, however, are less forgiving than pistons and give fewer warnings of trouble because of the closer tolerances to which they are made. This is why regular power checks (once a week) are carried out on them to keep an eye on their health. The other difference is that damage to a piston engine caused by mishandling tends to affect you, straight away, whereas that in a turbine tends to affect others down the line. In a turbine-engined helicopter, power used is indicated by the *torquemeter*.

Apart from sympathetic handling, the greatest factor in preserving engine life is temperature and its rate of change. Over and under leaning are detrimental to engine life, and sudden cooling is as bad as overheating—chopping the throttle at height causes the cylinder head to shrink and crack with the obvious results—the thermal shock and extra lead is worth about $100 in terms of lost engine life. In other words, don't let the plane drive the engine, but rather cut power to the point where it's doing a little work. This is because the reduced power lowers the pressure that keeps piston rings against the wall of the cylinder, so oil leaks past and glazes on the hot surfaces, degrading any sealing obtained by compression. The only way to get rid of the glaze is by honing, which means a top-end overhaul. For the same reasons, a new (or rebuilt) engine should be run in hard, not less than 65% power, but preferably 70-75%, according to Textron Lycoming, so

the rings are forced to seat in properly. This means not flying above 8000 feet density altitude for non-turbocharged engines. Richer mixtures are important as well. Also, open the engine compartment after shutting down on a hot day, as many external components will have suddenly lost their cooling. With some turbine engines (like on the AStar), you have to keep a track of the number of times you fluctuate between a range of power settings because of the heat stress.

When levelling in the cruise, the combination of increased speed and throttling back cools the engine rapidly, so close the cowl flaps beforehand. Don't use the cowl flaps as airbrakes, either, but to warm the engine after starting and to cool it after landing (allow temperatures to stabilise before shutting down, especially with turbochargers).

One point with low power settings when it's very cold is that the engine may not warm up properly and water forming from the combustion process may not evaporate, so oil won't lubricate properly.

Although many flight manuals state that as soon an engine is running without stuttering it's safe to use it to its fullest extent, try warming up for a few minutes before applying any load, at least until you get a positive indication on the oil temperature (and pressure) gauges. This ensures a film of oil over all parts.

Even better, warm it before you start it, because the insides contract at different rates – in really cold weather the engine block may have the grip of death on the pistons and strain them when the starter is turned. Equally important is not letting it idle when cold, as you need it to be fast enough to create a splash of oil inside (1,000 RPM is fine).

After flight, many engines have a *rundown period* which must be strictly observed if you want to keep it for any length of time. As engines get smaller relative to power output, they have to work harder. Also, in turbines, there are no heavy areas to act as heat sinks, like the fins on a piston engine, which results in localised hotspots which may deform, but are safe if cooled properly, with the help of circulating oil inside the engine (75% of the air taken into a turbine is for cooling purposes). If you shut down too quickly, the oil no longer circulates, which means that it may carbonise on the still-hot surfaces, and build up enough to prevent the relevant parts from turning. This coking up could sieze the engine within 50 hours or less.

If the starter light remains on after you release the starter button on a piston engine, you should shut it down, as it indicates that the starter is still engaged with the engine and is being driven by it.

Lubrication

Friction can be quite handy, but not inside an engine. Without some way of making the various surfaces rub smoothly against each other, they would get hot, and suffer from scoring damage.

Oil actually does many things, including *cooling, cushioning, flushing, lubrication* and *sealing.*

There are two main methods of lubrication, *wet sump* and *dry sump.*

The first is very simple, with the engine oil in a sump under the engine, in which the crankshaft and other moving parts rotate, splashing it all around (*splash and mist*).

Dry sump uses a tank outside the engine, and oil is *force fed* around under pressure where it is needed (although wet sumps systems have pumps, too). The *scavenge pump* (which pulls oil from the engine) has a greater capacity than the pressure pump, to make sure the tank gets filled properly. The *filter* will be between the engine and the scavenge pump. The *oil cooler* is between the scavenge pump and the reservoir.

Pumps are usually *mesh gear* types:

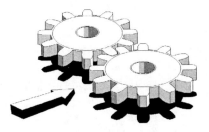

Oils come in various thicknesses, or *viscosities,* which measure resistance to flow. The lower the viscosity number, the thinner it is, so you would use 120 oil in Summer, 100 in Fall or Spring, 80 in Winter and 65 in the Arctic. To keep the oil thin, in the cold, one trick is to pour a few litres of petrol into the oil system just after closing down at night, so it is very thin in the morning and you can start the engine. By the time the oil has warmed up, the petrol has evaporated and you can carry on (but check your flight manual to see if this *Oil Dilution* is acceptable for your machine).

Oil is cooled by pumping it through an *oil cooler*, which is just like a radiator. An *oil filter* is used to trap any impurities, and the *pressure relief valve* is there to make sure it doesn't get too high (if the pressure increases, due to a blockage, maybe, the valve opens and dumps the oil back to the tank).

Chip Detectors are small magnets that attract slivers of metal suspended in the oil. Sometimes, they are connected to a warning panel in the cockpit in which a light glows if the sliver makes a circuit across the detector. It's always a good idea to be prepared to land straight away if you see a chip light come on, and some flight manuals say do so immediately. If the light is in a transmission system, keep it loaded, as unloading a disintegrating one has been known to make it worse (in a helicopter, make sure you land next to a pub; the engineers like it better).

An engine that is not used enough develops corrosion very quickly on the inside, and rust flakes, which are very abrasive, will circulate when the engine is started, which is why you have to change the oil even when you don't fly a lot. Another reason is an increased water content, which will have an acidic effect once it mixes with the byproducts of combustion, which is why you should just pull the propeller through several rotations if you cannot fly. The most wear takes place in the first seconds of a cold start, after the oil has been allowed to settle. Priming will wash whatever oil is left off the cylinder walls, so don't do too much, and maintain minimum RPM to let the oil circulate. The pressure will be high

just after starting, but will reduce to normal once the engine warms up.

Mineral oil has no additives and is used in new engines. *Detergent oil* has chemicals added to help with cleaning, etc., including keeping particles suspended. *Do not mix the two. Synthetic oils* have come from turbine oil development, but they have one drawback, in that the sludge tends to centrifuge out inside the dome of a constant speed propeller and make cycling a bit difficult. They also hold contaminants longer.

When flying, the oil temperature and pressure gauges work with each other (they are measured after the pump and before the engine). If the pressure is low, you can either expect the temperature to rise because it is working harder (PA 31), or reduce because there is less going over the temperature detector (Bell 407). Check your flight manual, but it also depends on whether oil is leaking, or whether the detectors are near each other.

With Hiller 12 and Bell 47 helicopters, the main rotor gearbox is lubricated with engine oil, because it is bolted directly to the top of the engine. An engine oil pressure problem with these machines (or any similar) is therefore quite serious.

After starting a cold engine, particularly in winter, you can allow the oil pressure not to rise for 30 seconds, because it may be too thick to get through the passages until it gets warm. Otherwise you should shut the engine down immediately.

If the temperature rises even with the cowl flaps open, the oil flow is blocked through the core of the cooler, which is probably why it is known as *coring*.

Fuel Supply

The simplest system is *gravity feed*, which needs the fuel cells to be above the engine to work properly (as with the Bell 47). Modern design requirements, however, mean that the fuel cells are in all manner of strange places, and come in many different shapes and sizes (together with a C of G system all of their own). Because of this, various methods are used to get the fuel from them to the engine, all involving *fuel pumps* and *filters*. Each engine will have its own pump, but there will also be an inline backup, just before the carburettor or actually at the fuel tank. Note that boost pumps are lubricated by the fuel they work on, so don't run them dry or you will burn them out.

A *fuel primer* is a small hand pump designed to put neat fuel directly into either the induction manifold (near the combustion chamber) or the inlet valve port before you start in the cold to promote the presence of fuel vapour that will ignite to start the engine (very rarely do you need to prime a warm engine). They are not there with fuel injection systems.

Fuel tanks will be *vented* to atmosphere, to prevent a vacuum forming inside the tanks as the fuel level is reduced. The vents might be in the fuel cap, or be an *overflow pipe* in the tank.

Although many fuel gauges are accurate, they should never be relied upon as the final guide to what you have. Reading the book *Free Fall*,

about the Gimli Glider is very instructive about this - a 757 had to make a dead stick landing at Gimli after running out of fuel in the cruise, due to a combination of circumstances, including misleading fuel gauges. Actually, the episode is also instructive with regard to CRM procedures.

Hydraulics

Liquids have minimal *compressibility*, meaning that, when pressure is applied, it will be pretty much taken up throughout the whole system. This makes it a useful way of transferring movement round corners and into strange places, as the forces produced by a hydraulic system can be very powerful indeed, which is why they are used in helicopters to reduce the forces that would otherwise be required to move the flying controls.

Keeping to the helicopter theme, some (such as the AStar) may have an *accumulator* instead of a second hydraulic system, to save weight. Aside from smoothing out fluctuations in hydraulic pressure, the accumulator's job is to store pressure that can be used for a short time if the main system fails. That is, it can be used in emergency. You can think of an accumulator as a shock absorber, since a valve opening in a highly pressurised system causes quite a shock to the lines. It is a cylinder in which a piston separates hydraulic pressure from air, which is pressurised on the ground.

Speaking of shocks to the system, hydraulic fluids are specially made to withstand high pressures and temperatures without vapourising, so make sure you use the proper stuff.

The basic system will have a *jack*, with a *control* to direct the fluid into whichever end of the jack is desired to move, a *pump* and a *reservoir*. The engine drives the pump, which moves the fluid out of the reservoir and applies pressure to it. When the jack gets as far as it can go, the increased pressure forces a *relief valve* open so the fluid can be dumped back to the reservoir. To keep everything clean there will also be a filter somewhere.

Some helicopters have a *rotor brake* that is hydraulically operated, but is nothing to do with the main system. It consists of a disk around the main drive shaft that is gripped with brake pads operated by the lever in the cockpit. There will be a range of rotor RPM that the brake must be operated within.

Undercarriages

These can be skids, skis, wheels or floats. Helicopters can also have inflatable floats that are used when over water – they are packed tightly inside a covering and are inflated when the pilot operates an air bottle inside the cabin. Wheeled helicopters typically taxi like aeroplanes, and rarely hover for long periods.

Landing gear is there to take the shock of landing, so it isn't transferred to the airframe. Of course, it also helps you get around on the ground. The *retractable* variety produces less drag in flight, at the expense of complication. As with transmissions, the system will have its own oil system, in this case hydraulics, and a backup system should it decide not to work. This can be operated manually, or by an air bottle.

Otherwise, landing gear is made up of *struts*, that are attached to the fuselage, and which have the wheels attached to them (in tailwheel aircraft, struts are fitted slightly ahead of the C of G – otherwise, they will be slightly behind). For shock absorption, an *oleo strut* contains a piston and a cylinder, moving together inside hydraulic fluid. There are holes in the piston to allow the fluid through and damp down the shocks. There may be nitrogen or dry air instead of fluid. Another function of a strut is to force the tyre on to the ground.

With wheels, there will be brakes, used for stopping, and sometimes steering (not recommended in a multi-engined aeroplane – use differential engine power instead). There will be a master cylinder for each brake pedal, a reservoir, the brakes themselves and connecting pipes. The toe pedal activates the master cylinder directly. A piston inside forces fluid along the lines to activate the brakes on the wheels.

Transmissions

This is how you get the power from the engines to the propellers or rotor blades. Because some engines work at higher speeds than other components, there will also be an element of reduction involved. Transmission systems will also have their own, self-contained oil supply.

In a helicopter, a *clutch* may be fitted between the engine and main rotor gearbox. A *centrifugal clutch* (as found on the Bell 47) is automatic and will have more effect as speed is increased so, if the engine stops, the blades are free to rotate. A friction or belt drive clutch (like in the

Enstrom) is manually operated by the pilot, which allows the engine to be warmed up without the blades rotating (safer for passengers, too). Some turbine helicopters have clutches, too, like the Gazelle. Others have a *freewheel* unit to let the blades spin without engine power.

The engine RPM and rotor RPM gauges are normally combined, with the needles superimposed on each other:

They are *split* in the above diagram.

Pneumatics

Compressed air is used for many purposes, such as operating the landing gear, doors, flaps, etc. Nitrogen will sometimes be used for backup pneumatics for landing gear.

Flight instruments are also operated by air, and for deicing boots.

Heating Systems

The simplest system uses a muff round the exhaust that makes air flowing through it warmer, on its way to the cabin. As a result, there is a danger of carbon monoxide poisoning if the exhaust is faulty.

Slightly more complicated is what can best be described as a flame in a tube, under some control, of course. These need a rundown period which

should be observed, otherwise carbon will form on the igniter and stop it from reigniting.

A turbine powered machine can use bleed air from the engine, which can be mixed with outside air to regulate the temperature.

Fire Detection

Fire has three elements—fuel, oxygen and the heat. Take one away and it stops. With dangerous goods, you can get fire from the reaction of flammable materials with an oxidising agent – you don't necessarily need a source of ignition.

A **Class A** fire is an ordinary one, that is, of normal combustible material on which water is most effective. A **Class B** fire is in a flammable liquid, such as oil or grease, where you would probably use a blanket. A **Class C** fire is electrical, for which you need a non-conducting extinguisher. For the latter two, you could use either CO_2 or Dry Powder (which ruins the avionics), but the fumes may be toxic, so you will need plenty of ventilation afterwards. You can use Halon on anything, if you're allowed to use it. A **Class D** covers other materials, like metals, that may burn if persuaded.

To help you identify the source, smoke associated with electrical fires is usually grey or tan and very irritating to the nose or eyes (it doesn't smell too good, either). Anything else (say from the heater) tends to be white, but you may get some black from upholstery.

If you think you have an electrical fire, it's no good just using the extinguisher, because you may be treating the symptom and not the cause, although there is a school of thought that advocates not using an extinguisher at all if you can possibly help it, due to the fumes and stuff you have to breathe in until you land. Whatever you do, transmit a Mayday before it's too late—you can always downgrade it afterwards. Bear in mind also that your first strike with your extinguisher is the best, because the contents and pressure decrease from then on.

Next, put an oxygen or smoke mask on, if you have one, then bring on *essential* electrics one at a time until the smoke appears again.

On the ground, engine fire drills may vary considerably between different types, and these will have to be memorized, but there are some general points that can be made. One is, before evacuating the aircraft, make sure the parking brake is off, so it can be moved somewhere safer if things get out of hand, always being aware that it could run off by itself, as well! If the fire has been caused by spilt fuel, has spread to the ground under the wing and the other engine has been started, taxi clear of the area (or more specifically, the fuel on the ground) before evacuation, keeping the fire on the downwind side. If the other engine has not been started, evacuate first, carrying out what drills you can.

If you can, use the radio to summon help, and take the extinguisher. Remember that human beings *en masse* need very different handling than when encountered singly.

In the air, initial shut down actions are similar everywhere—after performing vital actions from memory (e.g. identifying the source and all that), refer to the checklist to see if you haven't forgotten anything.

If the engine has been secured promptly, the fire should go out quickly after the fuel supply has been cut off. You will find, however, that structural failure of the wing will be imminent after about two minutes if the fire is uncontrolled, which is a sobering enough thought to make you commence emergency descent IMMEDIATELY, no matter how good it looks.

If you've got fire extinguishers in the engine bays, delay actuating them until the engine has been secured and you've no reason to suspect a false alarm; that is, unless you can actually see signs of a fire. In the cabin, whether in the air or on the ground, the priority is to get out, and as soon as possible, because if the flames don't get you, the fumes will. The only difference between the two situations is how quickly this can be done, and what you can do about it.

Heat sensors can be found in many engine compartments, with *smoke detectors* in cabins, toilets, or anywhere convenient. *Infra red* systems detect the light from flames or glowing metal (also used by railways to detect hot wheels). Some systems in small helicopters use a very thin wire, with another even thinner one inside, separated by a conductor that will vapourise with heat. When the two wires connect, a circuit is completed and a warning will go off. Sometimes, however, theses can short out, particularly on misty or otherwise humid days, and create false alarms.

If there isn't actually an extinguisher in the engine, at the very least there should be a *firewall shutoff*, operated from the cockpit, that will stop the flow in fuel and hydraulic lines (PA 31, etc). If there isn't one of those, try for a *fuel valve* that stops the flow to the engine (Bell 206).

An engine-based system will have CO_2 or Halon in cylinders.

The recommended extinguisher for wheel assemblies is dry powder.

Autopilots, etc

Those in large aeroplanes control attitudes in pitch, roll and yaw and are known as 3-axis. You will be able to maintain an altitude or heading, intercept and follow a radial or localiser and keep to a descent pattern. However, when following a radial, the results can often be unsettling to passengers, as the system chases the needle too much. A better solution is to use the heading bug and chase the needles yourself so you don't spill the coffee. ·

A flight management system takes the input from several sources and interfaces it with the autopilot, aside from storing commonly used coordinates and routes.

In helicopters, a *Stability Control Augmentation System* (SCAS) is used for short term control assistance, just to reduce the workload. It is found in many machines, particularly the Hughes 500, which uses a beep trim on the cyclic control. *Rate gyros* generate electrical impulses that control actuators that resist movement from where the controls should be. To stop it going out of control, there will be a limiter.

The *Automatic Flight Control System* (AFCS) is a step above, using a computer to memories control positions and keep them there.

Vibrations

In helicopters, vibrations come from many sources, not least the operation of the machine itself – there are many spinning parts that must be finely balanced. One good source is being downwind. Various components can induce vibrations in others, for example, a tail rotor can vary them in the main rotors. They may be felt as *lateral, longitudinal* or *vertical*, or a combination. One way of testing which types are affecting you is to sit in the cockpit and rest your wrist on your knee to see which way your hand moves.

Vibrations fall into three ranges:

- *Low*, with large amplitude, between 100-400 cycles per minute, generally associated with the main rotor. They can usually be felt through the cyclic. A *wumper* is one kick per revolution (one per). A vibration in the stick and fuselage is possibly from the rotors or the rotor support system, particularly friction dampers. If felt in the controls only, look in the linkages.

- *Medium*, 1,000-2000 cycles per minute, usually stemming from the tail rotor after improper rigging or imbalances.

- *High*, over 2000 cycles/minute, from the engine, usually.

Aircraft Husbandry

You can tell how well pilots treat their machines by the maintenance costs. For example, not using brakes excessively while taxying can save a lot of money in a large fleet. Similarly, whilst the manufacturers tell you that you can use the rotor brake on a helicopter all the time, do you really need to? They tell you that, certainly, but they want to sell you spare parts! Even something as small as caging the gyros before you turn them on can make a lot of difference long-term.

Aircraft should not be parked on soft or sloping ground, and suitable chocks should be placed under the main gear wheels of aeroplanes. They should be parked into wind whenever possible, with the nosewheel in line with the fore and aft axis.

Control locks are devices attached to flying controls (external) or control columns (internal), to stop them moving on the ground and protect them from gust and high wind damage. Together with covers, they should be used whenever convenient (especially when winds will be high, in which case consider picketing as well), and all doors, hatches and windows should be closed when the aircraft is left unattended.

Control stops are devices that restrict a control's range of operation.

Anti-collision lights should be switched on immediately before starting engines, but it is suggested (like the military) that this be done immediately the aircraft is occupied, always having due regard for the capabilities of your battery. Speaking of which, always leave the anti-col light switches on when leaving the aircraft, because that lets you know you've left the Master switch on.

Don't forget taxying procedures, such as not using the brakes too much, or using aircraft momentum when turning corners to save using the engine. Engine runups (like on power checks) should be done into wind for better engine cooling and least strain on the prop, and away from loose items on the ground, both to protect people behind and the prop, as the airflow around the tips will tend to pull bits of gravel etc towards it, and cause damage.

When tyres touch the runway, and have to spin rapidly in a short time, they can *creep* round the wheel rim. Aside from stressing the tyre, it can also force the valve assembly to one side, so it is usual to monitor creep by checking the alignment marks on the tyre that are placed on it when the tyre is fitted:

Lastly, let me mention oil cans, which come sealed so you need a special implement to open them. Actually, you can use a screwdriver, but whatever you use, don't bang it down on the lid, but gently prise it open. This stops you getting slivers of metal in the oil which may disagree with your engine

Instruments

Before we start:

The instrument experience you get on your basic licence course does not qualify you for proper instrument flying! You must learn to overcome many of your body's limitation to do it properly.

OK, having said that, flight instruments are in a common format, called a *T arrangement*:

I know it doesn't look like one, but ignore the bottom right and left for the moment. The *artificial horizon* is in the centre, because it is a *primary instrument* (it tells you which way is up), the *heading indicator* is below, *No 1 altimeter* at the top right, with the *vertical speed indicator* below, and the *airspeed indicator* is at the top left with the *turn coordinator* underneath.

A *primary* instrument is one which gives instant and constant readouts (also *direct*). A *secondary* instrument is one that you have to deduce things from, such as the altimeter increasing, telling you that the pitch must have changed (you might also say that the altimeter gives you an *indirect indication* of pitch attitude). The ASI and VSI also give indirect indications of pitch, and the HI and TC indicate bank. Note also that a primary instrument will tell you at what rate things are changing, but a secondary one will only indicate that change is taking place.

They are further grouped under the headings of *pitch*, *bank* and *power*.

Pitch

- *Artificial Horizon* (Attitude Indicator). The most important pitch instrument, because it gives direct, instantaneous readings.

- *Altimeter*. Although it indicates pitch indirectly, it is a primary pitch instrument.

- *Airspeed Indicator*. A secondary pitch instrument.

- *Vertical Speed Indicator* (VSI). A secondary pitch instrument.

Bank

- *Artificial Horizon* (Attitude Indicator). Also the most important bank instrument, for similar reasons to pitch.

- *Heading Indicator*. An indirect instrument, because if you change heading, pitch must be involved somewhere.

- *Turn Coordinator*. As it shows a rate of turn (3° per second for rate 1), it is an indirect indication of bank.

Power

Not in the traditional T, but you have to keep an eye on them anyway.

- *Engine RPM*. Direct indication of power. Turbines rotate so fast that the numbers are too large to make sense of, so percentages are used instead (that is, 100% means full power). In a helicopter, the engine and rotor RPM needles usually sit on top of each other in the same instrument, although they can be separate. In powered flight, the needles are joined; in autorotation, they are split. Both conditions have a range outside which the rotor needle should not go. A turbine helicopter will also have a

smaller gauge showing "gas producer" RPM.

- The *Manifold Pressure* gauge shows the pressure inside the inlet manifold, measured in inches. The theory is that the higher the pressure, the more the amount of fuel/air mixture that is potentially available. The equivalent of the manifold pressure gauge in a turbine helicopter is the *torquemeter*, usually expressed as a percentage. Although 100% is the usual maximum, it may often be increased for a few seconds (check flight manual).

When the engine is running, MAP is below atmospheric because of the pressure drop across the throttle plate (butterfly valve). As the throttle is closed, the pressure drop will increase, and MAP will fall. When the engine is stopped, the MAP will be atmospheric.

You keep MAP constant with altitude by opening the throttle. Power will increase because exhaust back pressure falls, improving scavenging.

- The *Cylinder Head Temperature* (CHT) gauge shows you the temperature of a selected cylinder in a piston engine, but not necessarily the hottest (it's usually a rear one in a horizontally opposed engine). Operating an engine at a higher than intended temperature will cause loss of power, excessive oil consumption and damage to the cylinders.

- Knowing the *Exhaust Gas Temperature* (EGT) is useful when leaning mixtures.

- *Airspeed Indicator.* A secondary power instrument - it changes with power application.

- *Angle of Attack Indicator.* This is usually of a vane or a probe fitted flush with the side of the aircraft to detect relative air flow, to give an indication of the angle of attack. It provides a basis for the stall warning system and helps to verify the aircraft attitude and speed.

Pitot-Static System

This consists of a series of pipes through which air flows to feed three common instruments on your panel, the *altimeter, airspeed indicator* and *vertical speed indicator.*

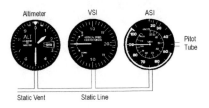

The system itself starts with a *pitot tube* (pronounced pee-toe) connected to the airspeed indicator (see below), to measure *dynamic pressure*, and a *static line* connected to all three (or four), to measure the *static pressure*, so called because it remains relatively static (it's actually the normal barometric pressure that decreases with height). The static lines are connected to *static ports* or *static vents* on the side of the machine, at right angles to the relative airflow. They may or may not be heated.

The altimeter therefore has two connections, the difference between them providing the basis of airspeed, assuming they are not blocked.

If the pitot tube and its drain get blocked, the airspeed indicator will read high in the climb, low in the descent and not change at all when airspeed varies. This is because only the static pressure is changing, so they are behaving like altimeters (a typical icing situation). If the drain hole remains open, however, the IAS will read zero, because there is no differential between static and dynamic pressures, due to the drain hole allowing the pressure in the lines to drop to atmospheric.

The pitot tube should be parallel to the relative airflow for best effect. It will be hot on most aircraft, as it needs to be protected against icing up, and a heating element will be switched on at all times, so be careful on your preflight and warn your passengers not to touch it.

Aircraft intended for IFR work will have an *alternate static source*, which takes its feed from *inside* the aircraft in case the main one gets blocked, either through ice, a bird strike, or whatever. If that is the case, the pressure read will be slightly lower, and will cause the airspeed and altimeter to read high.

The Altimeter

This is actually a barometer with the scale marked in feet rather than millibars. As you go up, the pressure will be less, which is the same effect as the pressure reducing at sea level. The altimeter, however, will be better sealed, so that air pressure in the cockpit doesn't affect it. The only pressure that should be there is that from the pitot-static system.

Inside the instrument are two *aneroid capsules* (vacuums), which are corrugated for strength and kept open with a large spring. They are very sensitive, and their movements as you go up and down are magnified by a linkage that connects them directly to the pointer. If they expand, as they would when you go up, the pointer increases the reading.

Outside, there is a small knob, which is linked to a subscale, visible through a small window. Rotating the knob causes the subscale to move and adjust the instrument to whatever altimeter setting you are flying through (see *Weather*).

Only in standard conditions will the true altitude be indicated. For example, when it is extremely cold, it will be a lot lower than indicated, so corrections must be applied (altitudes given with radar vectors from ATC are corrected already). If this is something you need to take note of, you could perhaps mark the corrections directly on to the approach chart, next to the heights they refer to.

The dials work in a similar way to a clock, in that the needles represent different scales:

The long, thin pointer indicates hundreds of feet and the short, wide one, thousands. A very thin one, maybe with an inverted triangle at the end, like the one above, shows feet in ten thousands.

Height is the vertical distance from a particular datum, usually in the case of aviation from the surface of an airfield (QFE is used more in Europe). *Altitude* is height above sea level. *Elevation* is the vertical distance of a point on the Earth's surface from mean sea level.

Indicated altitude is what is shown on the dial at the current altimeter setting. *Calibrated altitude* is the indicated altitude corrected for *instrument* and *position* error (see below). *True Altitude* is the actual one above mean sea level, taking the above errors into account, plus the air temperature and density.

You can calculate true altitude with a formula. First, subtract the ground elevation from the indicated altitude, and divide by 1,000 feet to get a single decimal number. Next, multiply that figure by the difference between the ISA temperature and the indicated one. Multiply that by 4 ft to get the amount to be *subtracted* from indicated altitude. Thus:

$$\frac{\text{Ind Alt-Elevation x OAT-ISA x 4 ft}}{1,000}$$

On the flight computer, put the PA against the OAT in the appropriate window and read the true altitude on the outer scale against the indicated one on the inner scale.

Pressure altitude is the height of a particular pressure setting, commonly 29.92 inches of mercury.

Density altitude is the pressure altitude corrected for non-standard temperature.

The *altimeter setting* (QNH) is the pressure at a point (or *station*, to be technical), corrected for temperature and reduced to mean sea level under standard conditions, so if you set it on your scale, you will see your height above mean sea level, or the airfield elevation when on the ground. In the latter case, to be serviceable, the altimeter should read within ± 50 feet (multiple altimeters should also be within ± 50 feet of each other, so they can misread by nearly 100' and still be OK).

Instrument and position errors will have been calculated by the manufacturers of the instrument and aircraft, respectively, and will be found in the flight manual. Position error arises because there is no perfect place to put the static ports (or the pitot tube, for that matter, when it comes to the ASI). Altimeters also suffer from *mechanical error*, due to misalignment in the linkages and gears, *temperature error*, particularly when cold, causing it to over-read, *elastic error* (hysteretic), a lag from stretching in the materials used in the capsule, found after large or rapid altitude changes, and *reversal error*, a momentary display in the wrong direction after an abrupt attitude change.

An *encoding altimeter* is used with a transponder in a Mode C system so that a height readout can be shown on a radar display.

Airspeed Indicator

To find airspeed, you need to compare the general pressure outside (the *static* pressure) with the *dynamic* pressure from the aircraft's movement through the air, so this instrument is connected to both the static and dynamic pressure systems.

It's similar to the altimeter inside, except that the capsule is fed directly with dynamic pressure, and its size varies in direct proportion to it. The needle, connected to the capsule, will read airspeed directly. Some aircraft, such as the Bell 407 helicopter, have a dampened needle, which will indicate the speed you *have been*, and not the speed you are at.

The instrument may be calibrated in *knots* or *mph*, that is, a rate of change of distance per unit of time. There are several variations, however:

- *Indicated airspeed* (IAS) is read directly, without corrections.

- *Calibrated airspeed* (CAS) is the IAS corrected for *instrument* and *position errors*, which are highest at low speeds (IAS and CAS will be about the same at speeds above cruise). It's known by older pilots as the *Rectified Airspeed* (RAS).

- *True Air Speed* (TAS) is the CAS corrected for altitude and temperature (remember its original calibration is based on the standard atmosphere). The slide rule part of the flight computer is used to calculate these, discussed below. On average, the TAS increases by 2% over the IAS for every 1,000 feet. Refer to the *Performance* chapter for a discussion on the effects of air density on TAS.

Various markings are quite useful if you don't have the flight manual to hand. The *green arc* covers the range of speeds for normal operations. The *yellow arc* is the *caution* range (that is, not to be used for long periods of time), and the *red line* is the speed not to be exceeded, V_{ne}. A *white arc*, on an aeroplane, is for flap operation.

If the pitot tube becomes blocked, you will see a gradual increase in speed with height (it's mostly a barometer, and therefore an altimeter inside). If the static is blocked, it will under-read in a climb and over-read in a descent (the best way to remember this is that it will always indicate more slowly than if it were working properly).

To find TAS, start with the CAS and Pressure Altitude. You will also need the temperature which, in an exam, may involve a conversion from Fahrenheit to Centigrade, and from miles per hour to knots. For example, given an altimeter setting of 30.40", an indicated altitude of 3450', an OAT of 41°F and an IAS of 138 mph, find the TAS in knots.

For the moment we will take CAS as 118 kts, having converted 138 mph to 120 kts and looked it up on an imaginary graph. If there isn't a graph, the question will contain the

information required. 41°F also converts to 5°C.

The PA is found in the usual way, remembering that 1" equals 1,000'. The difference between 29.92" and 30.40" is .48, or 480 feet, which gives 2970' when *subtracted* from 3450' (29.92 is the "higher" figure in terms of distance above ground).

The TAS is 122 kts, and the Density Altitude (out of interest) at 2500'.

Vertical Speed Indicator

There is a capsule inside this, too, but it is connected only to the static system. However, there is a *restrictor*, or *calibrated leak* between the inside and outside of the capsule that makes the pressure outside lag behind that on the inside.

The capsule is compressed or expanded one way or another and the "suitable linkage" transfers the movements to the dial to show climb or descent:

It is both a *trend* and a *rate* instrument, showing a direction of movement (up or down), and how fast you're doing it, in hundreds of feet per minute.

It suffers from *lag error*, which may last up to 6-8 seconds before the air inside and outside the capsule stabilizes, and *reversal error*, which occurs when abrupt changes cause movement briefly in the opposite direction.

Compass

The Earth has its own magnetic field, which resembles a doughnut, in that the lines of force are more or less parallel with the curvature of the Earth but increase their angle towards the Poles until they move vertically downwards in a circle surrounding the true pole:

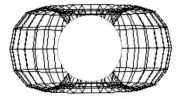

As the compass needle tries to follow the lines of force, you will find it trying to dip near the Poles, to a point where it is almost vertical and actually unreadable. This is why true tracks and headings are flown in those areas.

As the magnetic pole and lines of force do not coincide with either the true poles or lines of longitude, there is a system of accounting for *magnetic variation*, discussed in *Navigation*.

A direct reading compass has a pivoted magnet that is free to align itself with the horizontal component of the Earth's magnetic field. It must have certain properties to be able to do this, namely:

- *Horizontality.* The needle must dip as little as possible. This is done by making its centre of gravity lie below the pivot point, with pendulous magnets, which opposes the vertical component of the Earth's magnetic force (Z). Although there is still a residual dip, if it is less than 3° at mid-latitudes, it is OK. There is also a collar and sleeve assembly that stops it falling apart when inverted.

- *Sensitivity.* This can be improved by increasing the length and/or

the pole strength of the magnet. However, two short magnets will do just as well, and they can also be employed as the weights under the pivot point mentioned above. Pole strength can be increased by using special alloys. In addition, you could use a jewelled pivot to reduce friction, a suspension fluid which both lubricates it and reduces the effective weight of the whole assembly.

- *A periodicity.* The ability to settle quickly after a disturbance, which is helped by the suspension liquid. The two magnets employed above are also useful here, as they keep the mass of the assembly near the pivot, reducing inertia. Light alloys reduce inertia even more.

Being magnetic, the compass will be affected by all the fields generated by the aircraft itself, causing a phenomenon called *Deviation*, which is discussed in the *Navigation* chapter (see also *The Compass Swing*, below).

To try and eliminate errors, particularly magnetic dip, a *remote indicating gyrocompass* may be used, which is *slaved* to a DGI (see below). The *master unit* is mounted near the rear of the aircraft, so it is removed from as much influence as possible (hence the term *remote*). It contains a gyroscope under the influence of a magnetic element.

E2B

A typical E2B compass, as used in most aircraft today, consists of a floating inverted bowl suspended on a pedestal in kerosene (for damping):

The bearings are marked on the outside of the bowl, and there are two parallel magnetized needles inside, suspended under the pivot point, as mentioned above.

The centre of gravity's position below the suspension point gives rise to errors when accelerating or turning, caused by *magnetic dip* (the reason for a suspended bowl in the first place) and inertia.

Acceleration Errors

These are caused by inertia on East-West headings. Because the C of G of the compass is under the pivot point, accelerating makes the bulk of the compass lag behind the machine and displace the C of G:

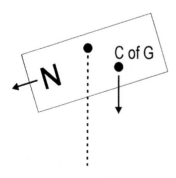

Because you are going East or West, the North bit of the compass is pointing to the side of the aircraft,

and the pivot and C of G are therefore side by side instead of being in line:

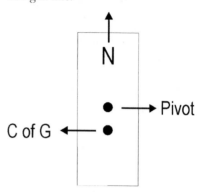

The pivot point (which is nearest to the Pole) is forced forward and the C of G goes backwards, making the North point spin further off line. The needle is forced clockwise to read less than 90° during the turn.

A deceleration would have the opposite effect. There is also a complementary effect from the vertical component of the Earth's magnetic force (Z) which imposes a turning force on the dipped end of the magnet – since the magnet can only turn by rotating about the pivot, the effect is created in the same way.

The watchword here is *A NDS – A celerate North, Decelerate South*, or *SA ND* in the Southern Hemisphere. Accelerations produce apparent turns towards the nearest Pole, and decelerations towards the Equator.

Turning Errors

These happen during turns through North or South – the compass *lags* on Northerly headings and *leads* on Southerly ones making it look as if you're turning slower through North and faster through South. Since a turn could be regarded as an acceleration, for the same reasons as mentioned above, as you bank facing North, the pivot point and C of G are still displaced in the same way, with the same effect on the North seeking pole, so the needle turns by itself in the same direction as the turn. At Rate 1, it may look like you are not turning at all - steeper than that, you could be going backwards! On Southerly headings, the machine will be turning against this inclination, so will read the opposite way, so don't straighten up until the compass has gone *past* the heading you want. The Z field effect is also active here, and is complementary.

To put it another way, during turns through the nearest Pole, the compass will be sluggish, so you need to roll out early. During turns through the furthest Pole, the compass will be lively, so roll out late. A displacement of the magnet in a clockwise direction viewed from above causes the compass to under-read, and vice versa.

You therefore get the most turning errors through North or South, and the most acceleration errors through East or West.

In view of the above, it follows that, before you start relying on the compass (either to navigate or align your HI), make sure you are in steady, level flight. Also, make turns gently, because the swirling of the fluid will keep the compass moving after you've stopped turning.

The Compass Swing
The influence of the aircraft on a compass is made up broadly of three components:

- *Hard iron*, or metal which can act as a permanent magnet.

- *Soft iron*, or metal which only produces a magnetic influence when affected by the Earth's field. This is because the lines of force flow more readily through metals then they do in air.

- *Electrical*. Current flowing through a conductor always produces a magnetic field.

Even though modern designs reduce aircraft effects as much as possible, there are always residuals to resolve. This is done by measuring the effects on the aircraft's compass against a Master compass, and introducing fields of equal magnitude but opposite polarity deliberately to counteract them.

Airfields and maintenance areas have clear areas in which this can be done. The aircraft is taxied there and everything that would be used in flight turned on. Then the aircraft compass is compared against a *landing compass* on several headings, in a *correcting swing*. Then the deviations are reduced by adjusting the magnets inside the compass and a *calibration swing* is done to see what deviations are left. These figures are written down on the deviation card.

Gyroscopes

Typically, three cockpit instruments are under gyroscopic influence. These are the *attitude indicator* (artificial horizon), *heading indicator* (DGI) and *turn indicator/coordinator*.

A gyroscope is a rotating mass on an axis, which may be *vertical* or *horizontal*. The spinning allows the gyro to maintain its own position in space (*rigidity*), regardless of whatever it is attached to is doing. In other words, it resists attempts to displace it from its position. If you attached one to a camera, for example, and used the camera in a helicopter, the helicopter could be bumping around all over the place due to wind or pilot input, and the camera would not move from where the operator put it. The same principle is used with the instruments mentioned above, as we shall shortly see.

Another property gyroscopes have is *precession*, meaning that a force applied to the spinning mass is felt 90° away from where it is applied, in the direction of rotation:

Force applied here

Is felt here

The control inputs on a helicopter have to allow for this, because the rotor disc is nothing but a large gyro – even though you make an input for forward flight, the actual movement applied to the rotor head is done several degrees beforehand. A more mundane example comes from riding a bicycle – when you apply a force to turn one way or another, it is done at the top of the wheels, but the turning movement appears 90° later, hence the turn.

If a primary precession is impeded for any reason, the impeding force will produce a *secondary precession* in the direction of the original force (as in the *turn and slip indicator*, below).

Gyroscopic instruments are made to spin through *suction* (heading and attitude indicator) or *electricity* (turn instruments). With the former, air is sucked out of the casing, and *vanes* (small bucket-shapes) on the gyro mass catch the movement and force it to go round (*vacuum* system). There might be a pump, or a venturi system (on older aircraft) to reduce the pressure. Since the venturi system depends on a tube aligned with the airflow outside the aircraft, it is only effective above about 90 kts, and therefore not good enough for IFR. The suction gauge on the instrument panel is always part of the checklist before IFR flight to ensure you have enough for the instruments to work properly. The rest of the vacuum system has a pump driven by the engine, a relief valve, an air filter, and enough tubing for the connections.

During startup checks, pull and hold any erection or caging knobs *before* turning the power on, as the parts inside can clash against each other as they spin up (just one of those little things a pilot can do to save long-term maintenance costs).

Artificial Horizon

Otherwise known as the *attitude indicator*, this instrument represents the natural horizon and indicates the pitch and bank attitudes, that is, whether the nose is up or down, or the wings are level or not. It is gyro-driven, and the gyro is *vertically mounted* so the housing can rotate around the vertical axis.

The horizon bar is connected to the rear of the gyroscope frame and the housing with a pin, so when the housing moves, the bar stays rigid.

With all these rotating parts, there is bound to be some friction, which will cause some errors. Others include acceleration error, found during forward movement (as in a takeoff) where a false climb is indicated. Deceleration will show a false descent.

Heading Indicator (DGI)

This works in a similar way to the artificial horizon, except that the gyro is horizontally mounted. The casing turns around the gyro, which has a compass card mounted on it. It is also air driven.

Because of wander (real or apparent - see above), unless you have a *slaved* compass (meaning automatic), you

should align this instrument with the magnetic compass every 15 minutes or so, remembering, of course, to do it in level, unaccelerated flight. You may get erroneous readings if the aircraft adopts unusual attitudes.

Turn Coordinator

This is actually a combination of two instruments, one power driven, and the other not. A small aircraft tilts to indicate whether you are banking, so it is a useful backup to the artificial horizon, especially since the gyro is electrically operated and not affected if the suction system fails (although it gives you a rudimentary indication of bank, turns without the other instruments are done with timing). The instrument is sensitive to yaw and roll, because the gyro's axis is tilted upwards by about 30-35°.

When the wings in the little aircraft hit one of the lower marks you are in a Rate 1 turn, which takes two minutes to go through 360°, making 3° per second (you can also add 7 kts to 10% of your airspeed to get a rough guide to the bank angle).

Underneath is a ball in a clear tube containing fluid, for damping purposes, called an *inclinometer*.

It is subject to gravity and centrifugal force, and will be thrown one way or another if the aircraft is not in a coordinated turn. In a *slip*, the rate of turn is too slow for the bank, so the centrifugal force will be less, and the ball will not be thrown out so much.

It will therefore be on the *inside* of the turn (decrease the angle or increase the rate to correct, or both):

In a *skid*, the turn is too fast, so more centrifugal force causes the ball to be displaced more, to the *outside* of the turn:

Correction is the opposite of the slip, above.

The turn indicator's gyro is laterally mounted, so it can tilt about the longitudinal axis. A linkage joins it to the pointer, and there is a restraining spring between it and the instrument case. There are mechanical stops to keep it from going more than 45° either side of the centre. As it is electrically driven, it is not affected by suction failure. Although it gives you a rudimentary indication of bank, turns without the other instruments are done with timing.

Rate of turn indications are only accurate at the speed for which the instrument has been calibrated, though these are not serious (around 5%). The angle of bank to obtain a given rate of turn increases with TAS, but you shouldn't need to make any calculations – the instrument reads correctly automatically. If the gyro rotates too slowly, the device will have less inertia and be less rigid, so it will tilt less and indicate a slower rate of turn than you are actually doing.

A Question

1. You've just lost all your electrics, and you are left with a map and the E2B compass. Can they be relied on to get you home?

An Answer

1. You can rely on the map, but the compass will have been swung with the electrics on, so all the local magnetic fields will have changed with them off. You can therefore expect large deviation errors.

Weather

Around the Earth is a collection of gases, called the *atmosphere*. 21% of it, luckily for us, is oxygen, but 78% is nitrogen, with 1% of odds and ends, like argon, that need not concern us here, plus bits of dust and the odd pollutant. What is important, however, is varying amounts of water vapour which will produce clouds. Because it weighs five-eighths of an equivalent amount of dry air, it will also reduce your engine's punch, but that's the subject of another chapter. The nitrogen, as an *inert gas*, is there to keep the amount of oxygen down, since it is actually quite corrosive.

The atmosphere is split into four concentric gaseous areas. Starting from the bottom, these are the *troposphere, stratosphere, mesosphere* and *thermosphere*, although the last two are not important right now. The first two are, however, and the boundary between them is called the *tropopause*, a freezing layer of dry air. Its height over the Equator is around 60,000 feet, more than it is at the Poles (35,000), because the air is warmer there and has expanded, taking the tropopause with it.

So, underneath the tropopause is the *troposphere*, and above it is the *stratosphere*, where the temperature remains relatively constant with height – it *decreases* with height in the troposphere, which is where weather happens (temperature stops decreasing at the tropopause). The troposphere contains more than 80% of the mass of the atmosphere.

Although the air gets thinner the higher you get, the proportions of the gases making it up stay the same, because of the constant mixing. If the air wasn't continually being stirred up, the heavier gases would sink to the lower levels.

The atmosphere will be wetter or drier, warmer or colder, or denser or lighter in different areas. The key words here are therefore *humidity, temperature, pressure, density* and *radiation*, as it behaves like any other gas, and obeys all the physical laws, such as expanding when heated, etc. Temperature, pressure and humidity all affect density, which ultimately affects aircraft performance.

The *climate* of any area is its average weather. Weather is what happens when the atmosphere is affected by heat, pressure, wind and moisture, but heat has arguably the most effect, since changes in weather occur when temperature changes.

Heat arises from the Sun's rays passing through the atmosphere and being converted to longer wave radiation when they hit the ground. The darker the area that is hit, the more absorption takes place, and the more heat is generated. Thus, any heat in the Earth comes from the Earth itself, and only indirectly from the Sun. In other words, moisture in the atmosphere acts like the glass in a greenhouse, that lets short wave radiation in, and long waves out.

The Seasons

We get seasons because the Earth is not vertical in space – it is actually inclined at an angle of 23 ½ ° so that different areas are pointed towards the Sun in their turn, and do it for longer periods, hence Summer.

Pressure Patterns

A column of air above any point has weight, which is commonly measured in terms of *millibars* or *inches of mercury*, and called the *atmospheric pressure*. This value won't change with temperature, but the *air density* will, as it is the weight of the air contained inside a cubic foot. As it gets hotter, the air in the imaginary cube will expand, and overrun the boundaries, leaving less air inside.

To make sure that everyone works on the same page, a couple of typical

scientists went to a typical place on the South coast of England many years ago and measured the temperature and pressure, which turned out to be 1013.2 millibars (29.92" of mercury) and 15° Centigrade. This was adopted as the *standard atmosphere*, and now everyone who makes altimeters, or whatever, calibrates them with it so you don't fly at the wrong levels. The pressure actually works out to be around 15 lbs per square inch, which equates to 20 tons on the average person. In short, *ISA is a standard that provides universal values of temperature, pressure, density and lapse rate, by which others can be compared.*

In the standard atmosphere, ½ sea level pressure is obtained at 18,000', 1/3 at 27,500' and ¼ at 33,700'.

Another quality of a column of air is that it gets cooler with height, as mentioned above, so the standard atmosphere is also taken as decreasing at 1.98°C per 1,000 feet, called the *standard lapse rate*, which is really an average, used for convenience. The *sea level pressure* on which it is based relates 1 inch of mercury to 1,000 feet of altitude, so you would expect to see an altimeter read 1,000 feet less if you set it to 28.92. 1 millibar is equal to 30 feet.

If you were faced with a question that asks you to compare an actual temperature at a height with the ISA standard, first of all, find out what the ISA temperature *should* be. For example, at 12,000 feet, it would be 15° minus the height times the lapse rate, in this case 12 x 1.98, so 15-23.76=-8.76, rounded to -9° for convenience. If the actual temperature were -7°, it would therefore be warmer than ISA, +2°.

Finally, the air gets thinner with height - at 18,000 feet, it is 50% of its density at sea level. This also means less oxygen, and difficulty in breathing, but this is covered elsewhere. Thus, the actual pressure at a given place depends on its height, and the temperature and density at that point (see *Density Altitude*, below).

Station pressure is the atmospheric pressure at a particular place. Several of these are taken, converted to sea level pressure and marked on a map, with the ones that are equal connected up. The lines that join the dots are called *isobars* (*iso* is Greek for *same*), and will be 4 mb apart.

The closer the isobars are together, the more the millibars drop per mile and the more severe the *pressure gradient* will be, so you get stronger winds (air moves from high to low pressure). Isobars are like contours on a map, and make common patterns, two of which are the *low* or *high*, the common names for *cyclone* and *anti-cyclone*, respectively (this has nothing to do with the cyclones that always seem to do severe damage to trailer parks. Another name for a low is a *depression*).

Three arrows (see below) are the traditional way of displaying the wind direction. The numbers in the middle of each figure are typical pressures (the exact position of a system will be marked by an X).

The arrows across the isobars are wind directions after *Coriolis Force* has taken effect (see below).

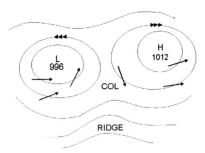

The patterns are round, being circles of isobars, and difference between them is simple; a high pressure area has most pressure in the middle, whilst a low has the least. In other words, the air in a high is descending (and diverging), and that in a low is ascending (and converging), so the weather is settled in the former and tends to instability in the latter, since upwards movement implies clouds forming. You might often see a *secondary low*, which is a smaller one inside a larger system, in which the weather is more intense as it feeds on its bigger brother, although the winds will be lighter between them. However, just because cloud is mostly absent in a high, don't expect clear skies, as the descending air might trap haze or smoke, leading to a phenomenon called *anticyclonic gloom* near industrial areas. Sometimes, areas of layer cloud may also get caught if they are below an *inversion*.

As an example of the influence a high pressure area can have, a strong one commonly sits over Eastern Canada in late Spring because Hudson Bay is still frozen. It is very good at stopping the movement of other systems and weather.

Other patterns are the *trough*, which is a longish area of low pressure, like a valley, with V-shaped isobars, which will be found between highs, and its opposite number, the *ridge*, found between two lows. A *col* is a neutral area between highs and lows.

A *complex low* is one with several fronts and air masses overlapping each other. When asked questions, try to create a 3D image that will show you which fronts and air masses are on top or below to get the sort of weather on the ground.

Air moves clockwise round a high and anticlockwise round a low, because the Earth is spinning, and air is deflected because of it. The Earth moves faster at the equator than it does at the Poles - if you threw something from the North Pole to the Equator, progressively more of the Earth's surface would pass under its track, giving the illusion of curving to the right (West) as it lags behind. If you threw whatever it was the other way, it would be advancing on the track and "moving East". This is called in some places the *Coriolis effect*, but technically is *Geostrophic Force*.

Thus, in the Northern Hemisphere, air coming from the South is deflected East, and West if coming from the North, which accounts for the anticlockwise movement, so, according to Professor *Buys Ballot's law*, if you stand with your back to the wind in the Northern hemisphere, the low pressure will be on your left. The implication of this is that, if you fly towards lower pressure, you will be drifting to starboard as the wind is from the left (common exam question, but worded differently). It's the opposite way round in the anticyclone, and in the Southern Hemisphere.

At about 2,000 feet, air movement is parallel to the isobars, but, as you descend, friction with trees, rocks, etc will slow it down by about 10 kts, which will lessen the coriolis effect and give you an effective change of wind direction to the left, known as *backing* (an increase to the right is called *veering*). Air moving round a low will therefore tend towards the centre and contribute towards the lifting characteristic.

Inside a high, though, air movement (i.e. winds), will tend to increase with the help of centrifugal force, other things being equal, but this is offset by the pressure gradient in a low being much steeper, creating stronger winds anyway.

Over the sea, the effect will be less, giving about 10° difference in direction, as opposed to the 30° you can expect over land. If the winds are high, you could get into a stall on landing as you encounter *windshear*, of which more later.

Wind

The Earth is heated unevenly. Air at the Equator becomes warmer than it does at the Poles, so it expands upwards around the middle of the Earth and contracts down to the surface at the Poles. This general trend gives rise to regular patterns of air movement, in the shape of winds that were well known to navigators on the high seas, such as the *trade winds,* caused by the Coriolis effect, which causes air to accumulate (for example) in an Easterly direction at around 30° of latitude in a general area of high pressure right round the Earth. Out of that high pressure, some air flows to the South West, and some to the North East. Air from the Poles, settling down and flowing South, creates weather fronts (see below) when it meets the warmer air.

Navigators also had to deal with the *doldrums*, which are areas of complete calm either side of the equator, where the only movement of air is up (it slides North and South once it hits the tropopause). Air gets sucked in from just outside, causing the wind. The doldrums move with the Sun according to season.

Wind is expressed as a velocity, so it needs direction and speed to fit the definition. It always comes *from* somewhere, expressed as a true bearing in weather reports (magnetic from the Tower), so a Southerly wind is from 180°. The speed is mostly in knots, or nautical miles per hour, as if you didn't know already. Wind speed is measured with an *anemometer*, while direction is measured with a *wind vane*.

turbulent eddies behind it, resulting in *gusts* and *lulls* as the speed varies.

Land and sea breezes arise out of a temperature difference between land and sea areas. When the land is warmer than the sea, the space left by the rising air over it is filled with more coming from over the water, producing a *sea breeze* (in fact, a relatively high pressure is created at about 1000 feet over land. With lower pressure at the same height over the water, there will be air movement towards the sea, which will subside to come back towards the land). At night, the process is reversed to get a *land breeze*.

Cool air on a slope will flow down, because it is more dense, and therefore more subject to gravity, causing a *katabatic* wind. It's the same effect you get in a closed room on a cold day, where there is a draught even when nothing is open - the air next to the window is cooled,

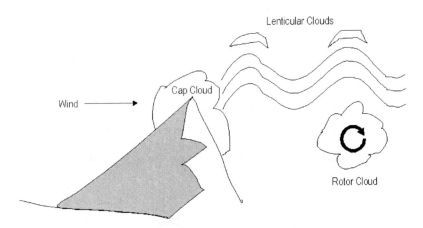

Lenticular Clouds

Wind

Cap Cloud

Rotor Cloud

Obstacles interfere with the wind in different ways. A forest acts as a large brush, slowing it down and mixing it up. It will tend to build up before an obstacle and create

and flows downwards. An *anabatic* wind flows up a hill.

Gusts are rapid changes of speed and direction that don't last long, whilst *squalls* do.

When a mountain range has an airflow greater than 20 knots blowing broadside on (within about 30°) and over it in stable conditions, *standing waves* may exist downwind, noticeable by turbulence and strong persistent up and down draughts:

The airflow follows the general shape of the surface and, flying into wind, you will experience a strong downdraught just before the ridge (the most dangerous bit, as it may be stronger than your climb capability) and an updraught just afterwards.

There are several miles between peaks and troughs of the waves, extending 10 or 20,000 feet above the range and up to 200 or 300 miles downwind.

You will see a cap cloud over the top of the range, creeping down the lee side (i.e. downwind), as a result of a downdraught. At the crest of each wave, there will be a *lenticular cloud*, with a *rotor cloud* downwind from each one. Rotors are always in circular motion, constantly forming and dissipating as water vapour is added and taken away. They are dangerous, and the most turbulence will be found in them, or between them and the ground. Lenticular clouds tend to remain stationary and will produce airframe icing.

An aircraft affected by mountain waves can expect severe turbulence below any rotors, downdraughts that may be stronger than the rate of climb and greater than normal icing in associated clouds.

Although the effects, such as turbulence and up and down draughts reduce with height, at normal cruise altitudes, mountain waves are usually free from clear air turbulence, unless associated with jetstreams or thunderstorms.

Watch out for long-term variations in speed and pitch attitude in level cruise (the variations may be large). Use the autopilot height-lock to maintain altitude, but change power as well - bear in mind that at cruise height the margin between low and high speed limits can be relatively small. Near the ground in a mountain wave area, severe turbulence and windshear (see below) may be encountered. The quickest way out of turbulence is up, with the next best directly away from the range. Flying parallel to the range in an updraught, avoiding peaks, gives the most comfort.

Low level windshear is found under the anvil of a cumulonimbus.

Windshear

This is the name for airspeed changes over about 10 kts resulting from sudden horizontal or vertical changes in wind velocity—more severe examples will change not only airspeed, but vertical speed and aircraft attitude as well. Officially, it becomes dangerous when the variations cause enough displacement from your flight path for substantial corrective action; *severe* windshear causes airspeed changes greater than 15 kts, or vertical speed changes over 500 feet per minute. Expect it to occur mostly inside 1,000 feet AGL, where it is most critical, because you can't quickly build up airspeed—

remember the old saying; altitude is money in the bank, but speed is money in the pocket.

Although mostly associated with thunderstorms (see below), where you have the unpredictability of *microbursts* to contend with, it's also present with *wake vortices*, temperature inversions, mountain waves and the passage of fronts, not forgetting obstructions near the runway, and can occur over any size of area. You can even get it where rain is falling from a cumulus cloud, as the air is getting dense from the cooling, and will therefore fall quicker. It's not restricted to aeroplanes, either—helicopters can suffer from it above and below tree top level in forest clearings, when a backlash effect can convert headwind to tailwind.

All fronts are zones of windshear—the greater the temperature difference across them (over 10°C), the greater the changes will be. Warm fronts tend to have less than cold ones, but as they're slower moving, you catch it for longer. In general, the faster the front moves (say, over 30 kts), the more vigorous the weather associated with it; if it goes slower, the visibility will be worse, but you can still get windshear even then and always for up to an hour after its passage.

Warm air moving horizontally above cold air can produce turbulence where they join, as would be typical with an inversion. In a valley, in particular, when moving warm air hits a mountainside, it will be forced downwards, but will not be able to penetrate the cold air, so it is forced to move over the top of the cold air

in the valley bottom, so watch out on those cold, clear mornings.

The most significant effect of windshear is, of course, loss of airspeed at a critical moment, similar to an effect in mountain flying, where a wind reversal could result in none at all! You would typically get this with a downburst from a convective type cloud, where, initially, you get an increase in airspeed from the extra headwind, but if you don't anticipate the reverse to happen as you get to the other side, you will not be in a position to cope with the resulting loss. This has led to the classifications of *performance increasing* or *performance decreasing*. With the former, you get more airspeed and lift from either increased headwind or decreased tailwind, taking you above the glidepath – recovery involves reducing power and lowering the nose, with a *higher* power setting than before when re-established, or the aircraft will sink. The latter is the opposite, of course – recover by increasing power and setting it to *less* than the original when established.

The effects also depend on the aircraft and its situation, in that propeller driven types suffer less than jets, and light aircraft tend to be less vulnerable than heavy ones—those with a good power to weight ratio will come off best. The take-off leaves you most vulnerable because of the small scope for energy conversion, less amounts of excess engine power and the amount of drag from the gear and flaps, which is not to say that landing is that much better.

In extremely simple terms, where windshear is expected, you should have a little extra airspeed in hand; you can help with the following:

- On take-off, use the longest runway, less flap and more airspeed up to about 1,000 feet AGL, but watch your gradient and use about 10 kts more than usual. If shear is indicated by rapidly fluctuating airspeed and/or rate of climb or descent, apply full power and aim to achieve maximum lift and distance from the ground. Be prepared to make relatively harsh control movements and power changes, using full throttle if you have to—new engines are cheaper than new aircraft.

- Set prop RPM to maximum (to get flat pitch).

Windshear should be reported as soon as possible, for the benefit of others.

Microbursts

These are small, intense downdraughts that spread out in all directions when they reach the surface, commonly associated with thunderstorms. You are most likely to encounter them within 1,000' of the ground, that is, right on the approach. They are most dangerous where the vertical push converts to the horizontal, between the base of the microburst and the ground – you could get a vertical speed of over 6000 feet per minute and a horizontal one over 45 kts. The diameter will be up to 5 km, and the duration anything between 1-5 minutes, or more, though the maximum intensity will start on touching the ground and only last for a couple of minutes.

They are problematical because they involve a performance-increasing shear to start with, followed by a performance decreasing one. Being so close to the ground, you are likely to be taking off or landing, and therefore more vulnerable. The angle of attack reduces inside a downburst, because it changes with the relative airflow, so the nose should be placed into a *high* pitch attitude on entry, and reduced (quickly) on exit.

Microbursts are rarely isolated – if you meet one, watch for another.

Wake Turbulence

A by-product of lift behind every aircraft, (including helicopters) in forward flight, arising from induced drag, particularly severe from heavy machines, and worst at slow speeds, as on takeoff or landing. *Wake vortices* are horizontally concentrated whirlwinds streaming from the wingtips, from the separation point between high pressure below and low pressure above the wing. Air flowing over the top of the wing tends to flow inward due to the reduced pressure sucking it in, while that under the wing tends to flow outwards because it is of higher pressure and pushes outwards. Where the lower air curls over the wingtip, it combines with the upper air to form a counter clockwise flow.

The distance between the vortices will be about ¾of the wingspan or rotor disc.

The heavier and slower the aircraft, the more severe they will be, and flaps, etc. will only have a small effect in breaking them up, so even *clean* aircraft are dangerous. The effects become undetectable after a time, varying from a few seconds to a few minutes after the departure or arrival, although they have been detected at 20 minutes. Vortices are most hazardous to other aircraft during take-off, initial climb, final approach and landing, but you should be careful any time you are within 1,000 feet below and behind a heavy aircraft.

Although there is a danger of shockloading, the biggest problem is loss of control near the ground. You are safest if you keep above the approach and take-off path of the other aircraft, or land beyond its touchdown point (or lift off before its takeoff point) but, for general purposes, allow at least 3 minutes behind any greater than the Light category for the effects to disappear (but see the table below).

Wake generation begins when the nosewheel lifts off on take-off and continues until it touches down again after landing. Vortices (one from each wing) will drift downwind, at about 400-500 fpm for larger aircraft, levelling out at about 900 feet below the altitude at which they were generated. Eventually they expand to occupy an oval area about 1 wingspan high and 2 wide, one on each side of the aircraft.

Those from large aircraft tend to move away from one another so, on a calm day, the runway itself will remain free, depending on how near the runway edge the offending wings were. They will also drift with wind, so your landings and take-offs should occur *upwind* of moving

heavy aircraft, *before* the point of take-off and *after* the point of landing. Inside a vortex core, you could get roll rates as much as $80°$ per second and experience downdraughts of over 1500 feet per minute, so avoid them.

Aircraft are grouped for wake turbulence into three groups:

Category	ICAO and Flight Plan	UK
Heavy (H)	136 000 or greater	136 000 or greater
Medium (M)	<136 000 and >7000	< 136 000 and > 40 000
Small (S)	–	40 000 or less and >17 000
Light (L)	7000 or less	17 000 or less

Successive aircraft on finals
Although ATC will normally suggest an interval, the table below can be used as a guide, although there is never a guarantee you will not encounter wake turbulence, whatever separations are given:

Leading Aircraft	Following Aircraft	Min dist (miles)
Heavy	Heavy	4
	Medium	5
	Small	6
	Light	8
Medium	Medium*	3
	Small	4
	Light	6
Small	Med or Small	3
	Light	4

Note: If the leading medium is a B757, increase to 4 miles, as they are difficult to slow down and lose height with, and often fly steeper approaches. BV234, Puma, Super Puma, EH 101 and S61N helicopters are Small. Bell 212, Sikorsky S76 and smaller machines are Light.

Departing aircraft
Applies to IFR and VFR flights.

Same or parallel runways less than 760m apart (inc grass)

Leading	Follow	Departing From	Min space
Heavy	Med/S m/Lt	Same takeoff posn	2 mins
Medium/ Small	Light	Same takeoff posn	2 mins
Heavy	Med/S m/ Lt	Intermediate posn	3 mins
Medium/ Small	Light	Intermediate posn	3 mins

Runways with displaced landing thresholds where flight paths cross

Leadiing	Following	Min space
Heavy	Arrival	Med/Small/Light Dep 2 mins
Heavy	Departure	Med/Small/Light Arr 2 mins
Medium	Arrival	Light/Small Dep 2 mins
Medium	Departure	Light/Small Arr 2 mins

Crossing and diverging or parallel runways over 760m apart

Lead	Crossing Behind	Min Dist	Time Equiv
Hvy	Hvy/Med/Sm /Lt	4/5/6/8 miles	2/3/3/4 mins
Med	Med/Sm/Lt	3/4/6 miles	2/2/3 mins
Small	Med or Sm/Lt	¾ miles	2/2 mins

Opposite direction runways
There should be at least 2 minutes between a light, small or medium and a heavy, and between light and a small or medium within 760 m (a grass strip is a runway).

Helicopters

Rotor downwash is wake turbulence from helicopters, which is easy to forget when hovering near a runway threshold or parked aircraft with little wind (although it's quite useful when crop spraying). Otherwise, the effects are similar to fixed wing, in that you get vortices from each side of the rotor disc, but the lower operating speed means they are more concentrated. Downwash also creates dust storms and can lift even heavy objects into the air, instantly presenting *Foreign Object Damage* (FOD) hazards to engines, main and tail rotor blades (so don't bolt your FOD, it gives you ingestion!—old RAF joke, on which I hope there's no copyright). Plastic bags or packaging sheets are FOD, too. Generally speaking, the larger the helicopter, the greater the potential danger (obvious, really). Bell 212, Sikorsky S76 and smaller machines are *Light*, in terms of the above table, but size is not significant when *creating* vortices; use the table for comparison purposes when avoiding other types.

The Altimeter

This is simply an aneroid barometer calibrated in feet rather than millibars or inches of mercury (its inner workings are described fully in the *Instruments* chapter). It measures the pressure at a given altitude, which is subtracted from the pressure at sea level. The difference is converted to get a height readout.

You would be very lucky to hit the standard atmosphere more than, say, 25% of the time, so you need a means of adjusting any instruments based on it to cope with the differences. An altimeter has a *setting window* in which you can adjust the figures for the correct pressure on the ground by turning a knob on the front (this is actually part of a very important preflight check, where you make sure that if you turn the knob to the right, the height readings increase, and *vice versa*. You also need to check that the reading given coincides with the airfield elevation, ±50 feet, and that, if you've got two altimeters, they are within ± 50 feet of each other (in other words, they can misread by nearly 100 feet and still be useable).

If you didn't adjust your instruments, and were flying between different areas of air pressure, you would not be at the height you thought you were, which is not that much of a problem if everybody else uses the same setting, as your relative height to other aircraft would be maintained, but it wouldn't with respect to hard objects, that is, obstacles, such as mountains, television masts, etc.

As an example, if you were flying from high to low pressure, your altimeter would be overreading (from HIGH to LOW, your instrument is HIGH), so you would be lower than planned and liable for a nasty surprise. It's therefore much safer to be going the other way (that is, from LOW to HIGH, where your instrument is LOW).

You can check what the difference is with simple maths, using the figures given above of 1" being equal to 1,000 feet. Remember that an increase in pressure equals a decrease in altitude, so if you start with 29.92, then go to where it's 30.92, the

altimeter reading would be 1,000 feet *less*, even though the figures themselves increase.

To convert from inches to millibars, in case you have an old altimeter, start at 29.92 and find the difference between it and the current pressure. Multiply that by 3.4 and apply it to 1013.2. For example, if the current pressure is 30.02, that is, 1 above 29.92", add 3.4 mb and set 1016.6.

The standard atmosphere has a temperature element that also affects the altimeter. Remembering that, as we said above, air density decreases as it gets warmer, a point in your imaginary column of air above a station would be higher on a warm day than otherwise:

If, therefore, as is typical near the Rockies in Winter, the air is very much colder than standard, you will be lower than you should be (actually, the phrase above is still valid, in that going from HIGH temperature to LOW, your instruments will be HIGH). This is serious because, in low temperatures, combined with other effects caused by movement of wind over ridges, you could be as much as *3000 feet* below your projected altitude, which could really spoil your day.

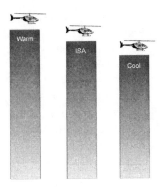

Another factor that arises from the above diagram is the creation of a wind purely from the temperature difference. The cooler column will have a lower pressure at altitude, and the warmer one will have a higher pressure, causing air movement to occur, from left to right, in this case. Applying Buys Ballot's law, low temperature is on the left in the Northern Hemisphere if you stand with your back to the wind.

When the surface temperature is well below ISA, correct your altitudes by:

Surface Temp (ISA)	Correction
–16°C to –30°C	+ 10%
–31°C to –50°C	+ 20%
–51°C or below	+ 25%

See also the *Instruments* chapter.

The *altimeter setting* is the station pressure reduced to mean sea level under ISA temperatures and standard lapse rates, that is, taking the elevation of the aerodrome into account, expressed in inches of mercury (Hg) in North America. It is what your altimeter must be set to when flying near aerodromes or other places that may issue it, because otherwise it will only tell you your height above the point you started from.

Its other importance is its use on weather maps to create isobars. To adjust station pressure for sea level, take the elevation and get its equivalent in inches. 900 feet would therefore be .9", which is *added* to whatever the reading is, because the sea is lower and would give lower figures anyway. MSL pressure is station pressure corrected to sea level using the average temperature over the last 12 hours.

Pressure altitude is the height within the standard atmosphere that you may find a given pressure (usually 29.92"), and is a favourite starting point for any calculations you may wish to make, certainly for performance, TAS, etc. To find it, get the local altimeter setting, find the difference between it and 29.92, convert it to feet (1"=1,000), then apply it the *opposite* side of 29.92 (you could get it from the altimeter itself, by placing 29.92 in the setting window, and reading the figures directly). For example, on an aerodrome 400 feet above sea level, with an altimeter setting of 29.72, your PA would be 600 feet, and where you would enter your chart (the altimeter setting is *below*, so your answer should go *above*). Again, you are adding because the sea is lower, and the figures ought to be higher.

In the exam, always write down 29.92 first, then subtract the altimeter setting and multiply by 1,000. If the answer is negative, take it away from the elevation. If it's positive, then add.

Density altitude is different altogether, and is your real altitude resulting from the effects of height, temperature and humidity, and is more to do with performance, as it is a figure that expresses where your machine thinks it is, as opposed to where it actually is – see under *Performance* in *Flight Planning*

Temperature

Believe it or not, most, if not all, of the Earth's heat comes from the Earth itself, that is, from below. The

Sun's rays do not produce heat (or light) until they hit something (that's why it's cold and dark in space), so the air will get warmed by *conduction* from the ground which has been heated up by them, known as *insolation*. Glider pilots and anyone who has done basic physics will know that lighter areas radiate heat better than dark areas do, so different parts of the earth will produce different amounts of heat, and *thermals*, which is what keeps gliders up in the air (for exam purposes, thermals are called *convection currents*, and they are just rising parcels of warm air). The ocean is always slower to warm up than the land, as are any marshy areas and forests. By comparison, rocks, roadways and pavements are very quick. As a general rule, the drier the better.

Since the Earth does not get hotter and hotter as the Sun shines on it, it follows that heat must be radiated away from the Earth. This explains the difference in temperatures between day and night, known as *diurnal variation*. The temperature begins to rise shortly after sunrise, and starts to fall mid-afternoon, carrying on through the night until the process starts again. This is less marked over water, which reacts more slowly. Changes over the sea will not be much more than 1°C.

Clouds will also absorb and reflect some energy from the Sun during the day, and act as a blanket overnight to stop heat being radiated from the Earth, further reducing diurnal differences.

There are two ways of measuring temperature, *Fahrenheit* or *Centigrade*

(Celsius), and it's a real pain to convert between the two. The quick and easy way is to use a flight computer, but here are the calculations for people who want to show off:

F - C $Tc = (5/9)*(Tf-32)$

C - F $Tf = ((9/5)*Tc)+32$

16°C is equal to 61°F, 20°C is 68°F and 30°C is 86°F, for quick conversions.

Given the standard of performance charts in the average flight manual, doubling the Celsius amount and adding 30 to get Fahrenheit, or subtracting 30 from Fahrenheit and dividing the remainder in half to get Celsius is probably good enough!

The Fahrenheit scale assumes that water freezes at 32°, and boils at 212°. Celsius starts at 0° and finishes at 100°, which is more logical, but the scale is coarser.

The freezing level is that where the temperature is 0°C.

If you apply an equal amount of heat to various substances, some will heat up quicker than others – the standard for comparison is that applied to water, which has a *specific heat* value of 1. If the same heat were applied to different amounts of the same substance, smaller quantities increase temperature more rapidly.

We have already seen that the standard reduction of temperature with height is 1.98°C per thousand feet. Where it remains constant, there is an *isothermal layer*. Where it increases (typical in anticyclonic conditions), you have an *inversion*, but the lapse process stops at the tropopause anyway. You may get slight turbulence flying through one.

Performance is affected by variations in temperature, and inversions will do so adversely. Large ones encountered shortly after take-off can seriously degrade climb performance, particularly when you're heavy. Even a small one in the upper levels can prevent you reaching a preferred cruising altitude. At lower levels, expect deteriorating visibility, as an inversion can prevent fog clearance for prolonged periods (to improve your chances of seeing the surface, fly higher above a mist layer). Another good reason for avoiding the top of an inversion is that all the industrial pollutants collect there, especially in the stubble burning season which may include incinerated pesticides.

However, the problem is that air is rarely dry, and cloud or water vapour will change the figures anyway.

Moisture

A given parcel of air can hold a certain amount of moisture at a certain temperature. This ability is increased as it gets warmer, and decreased as it gets colder. The *dewpoint* is the temperature at which it reaches 100% *saturation*, or the point at which water vapour begins the process of *condensation* into visible water droplets (the *condensation level*), so if the temperature and dewpoint at an airfield are the same, it will take very little incentive for clouds to form - the further apart they are, the less likely you are to get cloud, and therefore icing if the temperature is low enough. The warmer the wet air

is, the more likely you are to meet bad weather.

The *hygrometer* is one instrument used to measure how wet the air is, and it's very simple in the way it works. A human hair, which gets longer the moister it gets, is laid out against a calibrated scale of known humidities. A suitable linkage transmits its movements to show the *relative humidity*, which is how much moisture an air parcel is holding against what it could hold at that temperature or, in other words, the *percentage saturation*, which will *decrease* if the air gets warmer, as when subsiding in a high pressure area, because temperature is raised by compression, and can absorb more moisture (exam question). Relative humidity could change as a result of the air absorbing more moisture, say when moving over the sea, but it is more likely to change quickly through temperature changes, at least for our purposes. *Water added to air makes it moister, and less dense, and therefore more likely to rise.*

Mostly, air is made to reach its saturation point by force, such as being moved up the sides of mountains or over large areas of slower moving air (*large scale ascent*) and, if the conditions are right, cloud will form. In fact, the ways of cooling air are many. Where horizontal movement of air over a cooler surface is involved, it is called *advective*, as found over the praires when air is moving from East to West, and a frequent cause of fog.

The convection currents we have already met with cause *adiabatic* cooling by lifting air so it expands and cools, and *expansion cooling* is really the same thing caused by upslope movement or *mechanical turbulence*, another name for low level air being mixed and moved upward.

Radiation cooling tends to happen overnight with clear skies, when the Earth's heat radiates out into space. If the winds are light, just enough to stir things up (3-5 kts), fog will form, predictably enough called *radiation fog*. Night-time cooling is less over water than land, as water traps heat – this is called the *maritime effect*. It could also cause a *nocturnal inversion*.

What's known as the *cloud effect* reduces the results, because they absorb some radiation themselves. The *topographical effect* results from cold air getting trapped in valleys and lower areas.

The process of *sublimation* occurs when water vapour goes directly to the solid state (i.e. ice) without going through a liquid stage, or *vice versa*.

Anyhow, the lapse rate changes when we add moisture to the mix. The *Dry Adiabatic Lapse Rate* (DALR) is 3°C per 1,000 feet, and you use it to find the cloud base, after which you switch to the *Saturated Adiabatic Lapse Rate* (SALR) of 1.5°C per 1,000 feet (*dry*, in these circumstances, just means a relative humidity of less than 100%).

For example, your ground temperature is 10°C, and the dewpoint 7°C. To find the cloudbase and freezing level, take the DALR and divide it into the difference between the temperature and dewpoint, which in this case is 3, then divide that by the lapse rate (3), so your cloudbase would be at 1,000 feet. Then use the SALR from the cloudbase to count down to zero, so divide 7°C (the dewpoint) by 1.5 and add the converted number in thousands of feet to the cloudbase, to get 4,660 (from 4.66), so the freezing level would be at 5660 feet.

The reason for the difference in lapse rates is *latent heat*. The word means *undeveloped*, implying that an amount of heat is lurking in the background waiting to do something. Converting water from one state to the other requires energy, which originally comes from the Sun's rays as water is evaporated in the first place, and is stored with the vapour. While there, it is known as latent, and released when the water condenses, hence the lesser rate that ascending saturated air cools at, since the air becomes warmer as a result. Latent heat becomes involved when you change the form of a substance without changing its temperature. There can be so much heat released that flight in normally stable layer cloud can be bumpy, due to internal eddying.

Latent heat is the reason for the *Chinook*, a warm wind common in the lee of the Rockies around Calgary, which can raise the air temperature to 20° in the middle of Winter. Saturated air made to rise by the mountains cools at SALR, and when it descends on the other side, having dropped its moisture, it warms at twice that, i.e. DALR, so you get a dry, warm wind with clear skies. In France, it is called the *Mistral*, and is accepted as a defence in court for weird actions, as the air is also ionised. In the Alps, it is known as the *Fohn Effect*.

Stability

Remembering the previous mention of convection currents, you can see that air has vertical movement as well as horizontal. The less there is the more stable it is, and the less bumpy, because it tends to resist vertical motion.

A cold air mass moving over a warmer surface will be unstable because the lower layers will pick up moisture and temperature, and start rising (the moisture makes the air less dense). This will carry on into the night over the sea, as the water will keep its heat better than the land will. On the other hand, a warm air mass over a cold one will have its lower layers reduced in temperature, possibly as far as an inversion, which is about as stable as you can get.

Instability arises when air warmer than that surrounding it begins to rise, as it is bound to do, because it is less dense. It may have been lifted in the first place by *convection, convergence, mechanical turbulence, orographic* (over a geographic barrier, such as a mountain range) or *frontal* means. The warmer it is when it starts, the more energy the bubble has to keep going, but it's really the lapse rate that determines when it stops (well, OK, humidity counts as well). As it rises, air expands, and therefore cools, beginning to match the air around it, until it eventually cools off quicker than the surrounding air, and stops. Once it becomes saturated, though (and cloud forms), cooling slows down and allows the ascent to continue further.

If the lapse rate lies to the left of the DALR (that is, it is steeper – see the diagram below), air is unstable.

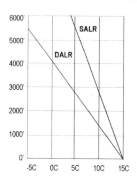

If it is between the DALR and SALR, it is *conditionally unstable*, meaning stable when dry, but not if saturated. To the right of the SALR (shallower), limited convection is possible, and to the right of the isothermal (i.e. vertical), you get total stability from an inversion.

If the lapse rate above the condensation level is greater than the SALR, the rising air actually gets warmer than that around it, which will give intense ascents and be instrumental in forming thunderstorms. Once this condition exists, all you need is a *trigger action* to cause condensation.

In summary, stable air can produce poor visibility at low levels, constant drizzle, light or calm winds with layer cloud, and no turbulence. Unstable air, on the hand, tends to be associated with good visibility, with heavy precipitation, heap type clouds, strong winds, turbulence and stormy weather.

Clouds

Cloud names were actually coined by an amateur meteorologist, Luke Howard, in 1803, who based them on the Latin words for *hair, heap, layer* and *rain-bearing* (*cirrus, cumulus, stratus* and *nimbus*) not to mention *middle* and *broken* (*alto* and *fracto*).

Clouds form in the first place because air contains water vapour, and because the air is cooled, causing the vapour to condense out at the *saturation point*. Air holds more vapour when it is warm, and a given amount can become saturated in two ways – you can either add more water to it, or reduce its temperature. The excess vapour changes from gas to liquid, with the droplets *coalescing* to form clouds. Most clouds arise from cooling – addition of water tends to happen when a dry air mass moves over a moist surface. Cooling occurs when air expands as it is forced upwards in various ways, already described under *Stability*. To remind you, though, they include:

- Uplift over a land mass

- Uplift over an air mass

- Convection currents

- Eddying (around the surface, or at the boundaries of two layers of air at different speeds)

- Wave action in the lee of mountains

- Uplift from a depression

Clouds affect surface heating by shielding the Earth's surface and absorbing the Sun's Rays, or acting like a blanket to keep the heat in at night. Those above the freezing level are largely ice crystals. Otherwise, there are two main types, *layer*, or *heap*, associated with stable and unstable conditions (see below), which might also be called *stratiform* or *cumuliform*, meaning horizontally or vertically developed, respectively.

There are a further three classifications based on the height of the cloud base, namely:

Low (Strato)
From sea level to about 6,500 feet, consisting mainly of water:

- *Stratus* (St), thin, uniform, low.

- *Stratocumulus* (Sc). Like stratus, but with small globules popping up here and there. Often formed in eddy currents at the boundary of air masses at different speeds.

- *Nimbostratus* (Ns for short), which is thick, dark, low rain cloud, typical in warm fronts.

Middle (Alto)
From 6,500-20,000 feet, made of water, ice, or super cooled water droplets, depending on temperature:

- *Altocumulus* (Ac) is similar to Sc (above), but higher.

- *Altostratus* (As), medium sheet greyish or bluish cloud of any thickness up to 10-12,000 feet.

High (Cirro)
From 20,000', made of ice crystals, so they have some transparency:

- *Cirrocumulus* (Cc) is high sheet cloud, made of small cloudlets (for want of a better word) which do not cast shadows. It looks like a *mackerel sky*.

- *Cirrostratus* (Cs) translucent high cloud, very delicate, made up of ice crystals. When in front of the Sun, you may see a halo round it.

- *Cirrus* (CIF) is a high and fibrous filament. It indicates that a warm front is around 200 nm away. Otherwise known as *Horse tails*, or *Mares' tails*.

The limits of each classification vary with latitude (with the troposphere). Low clouds do not have a prefix added to their name, medium cloud has *alto* in front and high clouds have *cirro*. *Nimbo* means rain-bearing.

A corona or halo around the Sun or Moon comes from light being refracted by moisture particles in medium-level cloud.

Other

Heap clouds, vertically developed, etc.:

- *Cumulus* (Cu), small amounts of heap cloud at low and medium levels, looking a bit like small balls of cotton wool:

It's actually convection cloud, which gives you a clue as to how it is made. Its outlines are much sharper in Winter than in Summer, but are ragged and woolly when decaying. In strong winds, you might see them in long lines, and you will get showers from larger ones. So-called "fair weather cumulus", typically seen on a nice Summer's day, are usually leftovers from dead Alto-Cu.

- *Cumulonimbus* is towering stormcloud, which may appear spontaneously over praires or be part of an advancing front associated with a depression:

"Towering" means up to 25,000 feet, and the anvil shape at the top is due to an inversion, which stops the cloud's ascent. They are mostly found around late afternoon. When they pass overhead, you may notice gusty winds and the altimeter overreading by 6-700 feet. There is a sleet centre around the freezing level, and severe up- and downdraughts, with cells of severe activity between them. Expect vast quantities of water in large drops, frozen or otherwise, but there are *supercooled water droplets* in the lower layers, drops that do not become solid when they reach zero degrees, but remain liquid. They do freeze, however, when they hit something, like a wing, where the liquid runs back and forms *clear ice* (see below).

- *Lenticular*, found at the crest of standing waves formed in the lee of mountain waves.

Precipitation

This comes from anything with *nimbo* in its name. It will be continuous from stratiform clouds, and intermittent from cumuliform clouds. *Virga* evaporates before reaching the ground - it comes from cumuliform cloud, and looks like streamers just below the cloud base.

Precipitation is the end result of a chain of events that starts with the cooling through ascent of a parcel of dirty moist air ("dirty" meaning that it contains microscopic particles that water can bind on to). Once the saturation point is reached, condensation occurs and droplets *coalesce* to fall out as rain, snow, or whatever, according to temperature.

Nitrogen oxides that come from lightning discharges can attract water droplets that can grow.

Hail & Sleet

Hail forms from large water droplets forced above the freezing level. Snowflakes are combined ice crystals which come from the freezing of water vapour without going through a liquid stage. Sleet is half-melted snow, that begins to unfreeze during descent below freezing level when it is quite high above the surface.

Ice usually forms on aircraft during flight in cloud, but it can happen in the clear.

Turbulence

This is found in cloud and clear air (that is *Clear Air Turbulence*, or CAT), and usually is a result of the friction from the mixing of air currents. It comes from various sources, such as convective, orographic, windshear and mechanical, and is reported as:

- *Light*, with small changes in height or attitude, near stratocumulus.

- *Moderate*, more severe, but you are still in control. A good

indicator is Cumulus-type clouds, which may also warn you about....

- *Severe*, with abrupt changes, and being temporarily out of control, indicated by Cumulonimbus and lenticular clouds, if there are many stacked on top of each other. Expect the latter when winds across mountain ranges are more than 40 kts.

- *Extreme*, which is impossible to control.

If turbulence is likely, use the turbulence speed in the flight manual, which will be rather less than normal. Mention it to the cabin crew and advise the passengers to return to, and/or remain in their seats, ensuring their seat belts/harnesses are securely fastened. Catering and other loose equipment should be stowed and secured until the risk has passed.

Thunderstorms

The airflow is greatly disturbed anywhere near a thunderstorm, usually noticeable by strong up and down draughts, together with heavy rain and lightning, or even tornadoes. Because of the inflow of warm air and the outflow of cold air, the gust front can extend up to 15-20 miles ahead of a moving storm. Avoid them even at the cost of diversion or an intermediate landing, but should this be impossible, there are certain things you can do.

It can be at least as dangerous up high as way down low—you can expect anything from lightning and turbulence to icing and hail, each with hazards of their own—lightning, for instance, could explode a fuel tank, and strikes can occur up to 20 nm from a storm cell. Not only that, even over baby ones near to larger storms, you will need at least 5000 feet clearance. Similarly, try not to fly underneath, either, or make steep turns. The currents inside a thunderstorm will easily be enough to suck in the average light aircraft, or spit it out.

A *squall line* is a series of storms, which Murphy's Law dictates will be right across your flight path (they will be too wide for a detour and too severe to penetrate):

They can appear anywhere the air is moist, but often ahead of cold fronts in late afternoon or early evening. They are the product of severe cold frontal conditions, in advance of the front, where it is nudging under the warm sector (watch for an acute bend in the isobars at the front, or low roll cloud across the advance).

Pressure usually falls rapidly as a thunderstorm approaches, then rises rapidly with the first gust. It returns to normal after it passes.

An embedded thunderstorm will have penetrated overlying bands of stratiform cloud on its way up.

To start a thunderstorm, you need moisture, a steep (unstable) lapse rate and a lifting, or trigger, agent, which could be *orographic, convective, frontal* or *nocturnal,* as occurs in the midwest plains after night time radiation from unusually moist air at height (of course, you could get two trigger actions, as when a front hits the Rockies). Here is a picture of one in the early stages of development:

A thunderstorm is actually a collection of several cloud cells in varying stages of development, with varying diameters.

During the *development stage* (see picture, above), several cumulus clouds will begin to merge, where the system consists mainly of updraughts, and will grow to around 4 miles wide at the base and 20,000 feet in height. Water droplets are merging as well to form larger raindrops, which get to be a hazard once they get above the freezing level and become supercooled (see *Icing,* below). When they are big enough, they will fall, and pull cold air down with them, which is where the downdraughts come from. So, rain at the surface is a good indication of the transition to ....

The *mature stage,* which is distinguished by the rainfall.

Updraughts won't get any faster, and the top of the cloud reaches the tropopause, where an inversion stops the ascent and strong winds produce the anvil shape. When downdraughts hit the ground, they spread out into a *gust front*, ahead of which are more, called *downbursts*, which may themselves contain *microbursts*. Expect lightning as well.

The characteristics of *dissipation* are downdraughts and disappearing cloud.

The different cells in a thunderstorm may be developing, maturing or dissipating at rates of their own, which could form their own trigger actions and make the storm self-perpetuating.

Approaching the area
Seat belts should be tightened, and loose articles stowed. One pilot should control the aircraft and the other should monitor the flight instruments. Select an altitude for penetration that will keep you clear of obstacles, and use the weather radar to select the safest track. Set the power for the recommended turbulence speed, adjust the trim and note its position, so any excessive changes from autopilot or mach trim can be quickly assessed. Height, mach, rate of climb or descent and airspeed locks should be disengaged but the yaw damper(s) should be on.

Switch on the pitot heaters, deicing, and continuous ignition system, where fitted. Disregard any beacons subject to interference, such as ADF (although tuning it to its lowest frequency will give you a primitive lightning detector). Turn the cockpit lighting fully on and lower crew seats

and visors to minimise the blinding effect of lightning flashes.

Within the Area
As the speed of vertical air currents may well exceed the capabilities of the aircraft, fly by attitude at the recommended turbulence speed and maintain your original heading—*do not* correct for altitude, except for obstacles; avoid harsh or excessive control movements, particularly with power, except to restore margins from stall warnings or high speed buffets. Do not be misled by conflicting indications on other instruments, and don't roll too much. If auto-trim variations are large, disengage the autopilot (movement of the mach trim, where it occurs, though, is necessary and desirable). Check the yaw-damper remains engaged. You might get temporary warnings (e.g. low oil pressure) from negative G, which may be ignored at your discretion.

Air Traffic Control Considerations
Obtain clearance from, or notify, ATC so they can separate you from others. If you can't, keep manoeuvres to a minimum, and inform them ASAP.

Take-off and Landing
Do not take off if a thunderstorm is overhead or within 5 nm. At destination, hold clear or divert.

Icing

Ice adversely affects performance, not only by adding weight, but also altering the shape of lift producing surfaces, which changes your stalling speed – autorotation in a helicopter could therefore be a lot more interesting than normal (the US Army found that half an inch on the leading edge reduces your lifting capacity by up to 50%, and increases drag by the same amount) – if your engine stops, you could *really* fall out of the sky!

On top of that, fuel could freeze in wing tanks, as could control surfaces, and slush picked up on take-off could stop the landing gear from operating, as well as flight instruments.

In fact, accident studies show that wings can stall 30% above normal stalling speed, and be undetected longitudinally. In addition, lateral control problems could lead to severe roll rates, up to ± *80° of roll.*

Zero degrees is when water becomes *capable* of freezing, from which you can infer that it doesn't necessarily do so. A *Supercooled Water Droplet* is one below freezing, but not frozen, because of the absence of *ice nuclei* to bind on to. *Hygroscopic nuclei* are needed for normal condensation of water, which could be almost anything microscopic floating around in the air, such as industrial haze or salt particles, but freezing requires ice nuclei.

When such a droplet strikes an airframe, however, just below 0°, some of it will freeze on impact, releasing latent heat and warming the remainder, which then flows back, turning into *clear ice*, which can gather without noticeable vibration (the airframe will act one giant ice nucleus, in other words). 1/80th part of a SWD will freeze on impact for each degree below zero. The worst place to penetrate cumuliform cloud

is between 0 to -10°C, where most SWDs are.

Rime ice comes from smaller SWDs well below 0° (actually from -10° to -40°C). It is opaque and granular and *moves forward* as it builds up on sharp surfaces like antennae. On a helicopter rotor blade, it is more likely to occur on the top rather than the leading edge. Below -40°C, you will encounter ice crystals only, which will not stick to the aircraft.

So—it's a good idea to avoid icing conditions but, in any case, having the equipment doesn't mean you can fly in icing conditions. On small twins it may just mean it produces no adverse effects on normal flight (though they might be nearly always overweight), and no-one could be bothered to take it off. Some aircraft are simply not happy in icing, even if the stuff is there (particularly true of older Barons and PA31s). Icing equipment is not certified if you are carrying deposits from ground operations or storage, so ensure that *all* frost, ice and snow is removed before you get airborne, if only because the aircraft systems don't get really under way till then

The trend now is towards a "clean aircraft concept" which, essentially, means that nothing should be on the outside of an aircraft that should not be there, except, perhaps, for deicing fluid, but even that is suspect.

All ice should be removed from critical areas before take-off, including *hoar frost* on the fuselage, because even a bad paint job will increase drag, which is relevant if you're heavy, and it will have a similar effect (hoar frost is a light frosty deposit that typically appears on a parked aircraft after a clear cold night. It can usually be seen through). These include control surfaces, rotors, propellers, stabilisers, control linkages, etc.

Although aircraft are different, expect icing to occur (in the engine intake area, anyway) whenever the OAT is below 4°C. Clear ice is found most often in cumulus clouds and unstable conditions between 0 and –10°C, and rime ice in stratiform clouds between –10 and –20°C (exam questions).

Ice is reported as:

- *Trace*, meaning slight, non-hazardous.

- *Light*, with occasional use of deicing equipment.

- *Moderate*, where use of above equipment is necessary.

- *Severe*, where the equipment is useless and you have to divert.

To keep out of trouble, before going, check that the freezing level is well above any minimum altitudes. Try to make sure the cloud tops are within reach as well, or that you have plenty of holes.

De-icing Fluids

The main types are what used to be known as AEA *(Association of European Airlines)* Type I (unthickened) with a high glycol content and low viscosity, and Type II (thickened) with a minimum glycol content of about 80% which, with a thickening agent (one or two teaspoons of corn flour), remains on surfaces for longer, but remember it has to blow off before you actually get airborne. The idea is to decrease the freezing point of water but, as the ice melts, the fluid mixes with the water, both diluting it and making it more runny (what's left after repeated applications to combat this is of an unknown concentration, and may refreeze quickly). Type III lies somewhere between the two. Type IV is similar to Type II, but with significantly longer *holdover times*. It is green, and needs care to provide uniform cover, especially over Type I fluid already there.

Type I fluids have good de-icing properties, but may refreeze - they *are for de-icing, not anti-icing*. Union Carbide *Ultra* fluid (i.e. Type IV) appears to increase holdover times by 1.5 over Type II and way more for Type I. The holdover time is how long the effects of the fluid should last – it can be affected by high winds or jet blasts damaging the fluid film, and temperature, humidity, etc.

General Precautions

Deposits must be swept from hinge areas and system intakes, and the sprays should not be directed to them, since the fluid may be further diluted by the melting ice it is designed to remove, and may refreeze. It may also cause smearing on cockpit windows and loss of vision during take-off.

Afterwards, confirm that flying and control surfaces are clear and move over their full range, and intake and drain holes are free of obstructions. Jet engine compressors should be rotated by hand to ensure they are not frozen. Check propeller spinners for trapped snow or moisture, which could subsequently refreeze and cause an imbalance. Don't forget the undercarriage.

Air Masses

A large body of air has the characteristics of its origin, particularly with regard to moisture and temperature. To acquire the uniform characteristics required to meet the classification, a mass of air has to stay in one place for several days in a more or less uniform place.

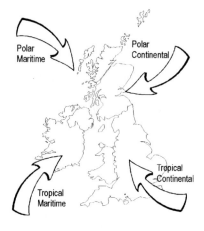

In UK, the main ones are:

*   and cold because it comes from polar land regions. It is only significant in Winter, but brings cloudless days. However, its track across the North Sea means it picks up moisture and will begin to resemble Polar Maritime, and there may be rain or snow in coastal regions, although cloud disperses inland.

*   *Tropical Continental*, dry and warm, during Summer only. The source is North Africa, and it is cooled from below as it tracks

North, so it stays stable, and will tend to poor visibility, despite cloudless skies. It stays dry because its journey is mostly over land.

*   *Polar Maritime*, which is cold, moist and stable at its origin, becomes warmer and more unstable as it comes South East across the North Atlantic (surface heating increases the temperature). It is moist because of surface evaporation, leading to convective cumuliform cloud. Any instability is increased during the day in Summer. In showers, visibility will be poor, but will improve as the air mass moves inland as moisture is lost.

*   *Tropical Maritime*, moist and warm, from tropical oceans. It picks up moisture on its travels, so relative humidity increases and the air mass becomes more stable as it moves North. In Winter, you can expect stratiform cloud and drizzle, with poor visibility remaining after cloud has been heated away by the warmer land mass. In Spring and early Summer, you will get advection fog in Western areas, particularly Cornwall and Devon.

They can be modified if they move over different areas.

Frontology

A front is a *line of discontinuity*, or a narrow transition zone between air masses where they are forced to mix, even though they don't want to. The difference is usually in temperature, but may be purely due to moisture content. Fronts are always associated with depressions, which are sometimes referred to as a *frontal wave*. Fronts will rotate (anticlockwise) around the low.

Warm tropical air could be forced over colder arctic air, for example, because it is less dense and, if moist, will form a typical cloud structure that we on the ground can use to tell when a front is coming. The name of a front, that is, *warm*, or *cold*, comes from whichever air mass is overtaking the other, whereas the type of weather you get is determined by the stability and moisture content of the warm air mass (exam question). The actual temperature is less important than its relationship to that of the surface it is passing over.

The *Polar Front* is an area where south- and north-westerly airstreams meet to form long series of depressions, starting off the Atlantic Coast of North America.

Frontogenesis is the term for the forming of a front, and *frontolysis* the one for its dissipation. The cold air mass does not move at a *stationary front*, and you get an *upper front* when very cold air is caught on the surface with the weather higher up.

The Warm Front
This exists where warm air overtakes a colder air mass and is forced upwards, meaning clouds. Its symbol on a weather map, resembling beads of sweat, is:

The *frontal slope* has a gradient of somewhere between 1:150 and 1:200, although the clouds themselves will be about 5 miles high, starting with Nimbostratus at more or less ground level, through alto-stratus to cirrostratus (when flying towards it, you would see the clouds the other way round, of course). Once you start seeing cirrus clouds, you know that a warm front is somewhere ahead, anywhere between 300-600 miles away, or nearly 24 hours at a typical speed of about 25 kts, so have an overnight kit if you have to wait it out (rain will typically be 200 miles ahead). You can use the typical slope figure to figure out the cloud base in front of the system. At 100 miles, it will be 2,640 feet, which comes from 1/200*100, making half a mile, multiplied by 5280 (feet).

Clouds will appear in this order as you fly towards a warm front – cirrus, cirrostratus, altostratus, nimbostratus and stratus.

The shallow slope ensures that whatever is coming will last some time, and you can expect the pressure to fall, the cloud to get lower, the wind to back and increase in speed, rising humidity, bad visibility, drizzle and rain, though not necessarily in that order.

The freezing level will be lower in front than behind, and the slope means that freezing rain will be falling on anything underneath (see diagram above). Once you see ice pellets, expect freezing rain next.

As the front passes, the rain will stop, then become drizzle under an overcast sky, and the wind will veer. As humidity rises to saturation point, visibility will be poor. You will then be in the *warm sector*, where conditions will be more settled for a few hours. For the exam, when asked to predict the future position of a warm front (and the type of weather), use the direction of the isobars in the warm sector, since the front moves parallel to them. Since warm air finds it hard to displace cold air, a warm front will move at

more pronounced. Look for this icicle-like symbol:

A cold front moves at about the speed of the wind perpendicular to it just above the friction level (i.e. about 2,000 feet)), but they are

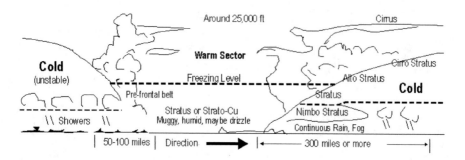

about 2/3 the speed of a cold front in the same conditions (see below).

After the warm sector comes......

The Cold front
This has a much steeper slope (1:50) and brisker activity, with more of a likelihood of thunderstorms. The rain becomes more showery and the wind veers more, to the West or Northwest. Pressure gets higher, and temperature and humidity decrease. In temperate climates, large amounts of Cu-nim are unusual at this point, but they are not over continental land areas. The rain belt is relatively small compared to the warm front, and visibility will improve markedly.

Expect questions on weather at a cold front, and after its passage (good vis, some turbulence – see below). Wind shifts will be usually

faster in Winter because the air is colder and exerts greater pressure. It's generally colder after its passage, and with less cloud, because pressure is greater to the West and less to the East, limiting the inflow of air.

The Occlusion
This marks the final stages of a frontal depression, and occurs because cold fronts move faster than warm fronts. When one catches up with the other, the warm sector (the bit between a warm and a cold front) is lifted from the ground altogether, leaving only one front on the surface (the rate depends on the temperature difference between the air masses). Here is its symbol:

The same naming convention applies, and you get more rain with a warm occlusion. You also get a quicker transition from warm to cold front weather. A cold occlusion is more or less the same as...

Trowal

A *Trough of Warm Air Aloft*, from a cold front catching up with a warm one.

Secondary Front

You get one of these when air just behind a cold front flows down and undercuts the warm air ahead. As it compresses it gets warmer and forms another front.

Visibility

Defined as the greatest horizontal distance a dark object (of known dimensions) can be seen and recognised against a light background. This, of course poses a problem at night, so night visibility really refers to how far you would be able to see in daylight.

Visibility may be reduced by fog, mist, cloud, precipitation, sea spray, smoke, sand, dust and industrial haze (you may think of others). The best visibility in haze is obtained when down-sun and up-moon.

Fog

This is essentially cloud at ground level, when you can't see more than 1,000 metres because of water droplets suspended in the air.

Radiation fog forms over land, preferably low-lying, when temperatures approach the dewpoint with very slight winds (2-8 kts), and where moisture is present. It doesn't

form over the sea, because the diurnal temperature variation is less, and it is often found in the early morning after a clear night, since its formation is most favoured by high relative humidity, light winds and clear skies. It usually clears quickly, once the Sun's heat gets to work, and will often get worse before it gets better. If the windspeed is too low, water cannot be held in suspension, so it falls out as dew. If it is too high, you will get low stratus. It will disperse with wind, heat, or a drier air mass.

Upslope fog forms from the cooling of rising air up slopes.

Advection fog arises from warm air flowing over a cold surface. It is not the same as radiation fog because air movement is involved, and the coolness does not arise from diurnal variations, but longer periods, as with the sea, where this type of fog is commonly found. Over land, it could arise from a warm, moist air mass flowing over cold ground.

Hill fog is low cloud covering high ground, which may or may not have contributed to its formation.

Frontal fog may be low cloud over high ground, or come from rain through unsaturated air beneath.

In *shallow fog*, you may be able to see the whole of the approach and/or runway lights from a considerable distance, even though reports from the aerodrome indicate fog. On descending into the fog layer, your visual reference is likely to drop rapidly, in extreme cases from the full length of the runway and approach lights to a very small segment. This may give the impression of pitching nose up, making you more likely to hit the ground after corrective movements. Be prepared for a missed approach whenever you have the slightest doubt about forward visibility. The minimum RVR for landing from a visual circuit is 1200m.

Mist

Essentially, thin fog, and the same definition applies, except that the visibility is over 1,000m.

Whiteout

Defined by the American Meteorological Society as:

"An atmospheric optical phenomenon of the polar regions in which the observer appears to be engulfed in a uniformly white glow".

That is, you can only see dark nearby objects – no shadows, horizon or clouds, and you lose depth perception. It occurs over unbroken snow cover beneath a uniformly overcast sky, when the light from both is about the same. Blowing snow doesn't help. It's particularly a problem if the ground is rising. Once you suspect whiteout, immediately climb or level off towards an area where you can see things properly.

Three common causes are:

- *Water Fog.* Thin clouds of supercooled water droplets contacting a cold snow surface.

- *Blowing Snow.* Winds over 20 kts picking up fine snow from the surface, diffusing sunlight.

- *Precipitation.* Small wind-driven snow crystals coming from low clouds having the sun shining

above them. Light is refracted and objects obscured.

Met Services & Information

ATIS (*Automatic Terminal Information Service*) is typically broadcast on a VOR frequency at major aerodromes (you can use it as an ID on instrument rides), although it may have its own channel. You should listen to it and take down the details before you contact ATC.

VOLMET is usually transmitted over HF for long-distance flights (North Atlantic and Arctic for Canada), but can be found elsewhere. It consists of long readouts of TAFs and METARs in a sequence, so if you miss the aerodrome you want, just wait for it to come round again. Many airfields have it available over the telephone.

Some codes (e.g. for wind velocity) use the same figures as the values being reported, so a wind from 280° at 15 knots is 28015KT. Otherwise, lettered abbreviations are used, as described below.

TAFs

These describe forecast conditions at an aerodrome for between 9 and 24 hours. The validity periods of many longer forecasts may not start for up to 8 hours after the time of origin, and the details only cover the last 18 hours. 9-hour TAFs are updated and re-issued every 3 hours, and 12- and 24-hour TAFs, every 6 hours, with amendments issued as and when necessary. They are not available for offshore operations. A TAF may be sub-divided into 2 or more self-contained parts by the abbreviation 'FM' (from) followed by the time UTC to the nearest hour, expressed as 2 figures. Many groups in METARs are also found in TAFs, but differences are noted below:

Message Type

TAF or TAF AMD, for amended. The amended forecast will have AMD inserted between TAF and the aerodrome identifier, and will cover the remainder of the validity period of the original forecast.

Station Identifier

4-letter ICAO sign for aerodrome.

Date and Time of Issue

A 6-digit code, with the date as the first two, then the time in UTC.

Validity Period

A METAR reports conditions at a specific time, but the TAF contains the date and time of origin, followed by the start and finish times of its validity period in whole hours UTC, e.g. TAF EGLL 130600Z (date and time of issue) 0716 (period of validity 0700 to 1600 hours UTC).

Winds

To the nearest 10°, in knots. 000000KT is calm, VRB means variable, less than 3 kts. Gusts are given in 2 digits. WS means windshear, when significant, with speed and direction at a height.

Horizontal Visibility

Only minimum visibility is forecast; RVR is not included.

Weather

If no significant weather is expected, this is omitted. After a change group, however, if the weather ceases to be significant, 'NSW' (no significant weather) will be inserted. A minus (-) means light, no sign is moderate, and + means heavy. It is described in 7 ways, such as SH (showers), DR for drifting, FZ (freezing), MI (shallow), BL for blowing and BC for batches.

FC=Funnel Cloud (Tornado), TS=Thunderstorm, DZ=Drizzle, FG=Fog (< 1 km), BR=Mist (> 1 km), GS=Small Hail, FU=Smoke, SS=Sandstorm, VA=Volcanic Ash, PO=Dust/Sand, RA=Rain, SG=Snow Grains, PL=Ice Pellets, IC=Ice Crystals, SA=Sand, SN=Snow, HZ=Haze, GR=Hail, DU=Dust, SQ=Squall, DS=Duststorm.

Cloud

Up to 4 cloud groups, in ascending order of bases, and *cumulative*, based on the amount of the sky covered, in *eighths*, or *oktas*. The *cloud ceiling* is the height of the first layer that is broken or overcast. The first group is the lowest individual layer; the second the next of more than 2 oktas and the third the next higher of more than 4 oktas. A group has 3 letters for the amount (FEW = 1 to 2 oktas, SCT, or scattered = 3 to 4 oktas; BKN, or broken, = 5 to 7 oktas, and OVC, or overcast = 8 oktas) and 3 for the height of the cloud base in hundreds of feet *above ground level*. For clear sky, expect SKC. VV means *vertical visibility* in hundreds of feet which, if you get it at all, means an *obscured ceiling*. CB means thunderstorms and is added as necessary. Clouds may cover the sky, but not conceal it if transparent, hence the term *opacity*.

Significant Changes

In addition to 'FM' and the time (see above) significant changes may be indicated by 'BECMG' (becoming) or 'TEMPO' (temporarily). 'BECMG' is followed by a four-figure group indicating the beginning and ending of the period when the change is expected. The change is expected to be permanent, and to occur at an unspecified time within it. 'TEMPO' will similarly be followed by a 4-figure time group, indicating temporary fluctuations. 'TEMPO' conditions are expected to last less than 1 hour in each instance, and in aggregate, less than half the period indicated.

Probability

Probability of a significant change, either 30 or 40%. The abbreviation 'PROB' will precede the percentage, followed by a time group, or a change and time group, e.g.:

    PROB 30 0507 0800FG BKN004

or

    PROB40 TEMPO 1416 TSRA
    BKN010CB

Example

    EGHH 0615 VRB06KT 9999
    SCT 030

was issued at Heathrow for the period 0600-1500, with variable wind at 6 kts, visibility more than 10 km and 3-4 oktas of cloud at 3000 above the airfield elevation.

METARs

Meteorological Aerodrome Reports are compiled half-hourly or hourly while the met office is open, about 15 minutes after observations are made.

Missing information may be indicated by oblique strokes.

They are *reports*, not *forecasts*, but you may see an outlook tagged on the end after the word TEND. It represents a two-hour period from the time of the observation. NOSIG means no significant changes are expected in the following two hours.

Message Type

METAR means a routine actual weather report. SPECI means a significant change off the hour.

Station Identifier

4-letter ICAO sign for aerodrome, and observation time in UTC.

Winds

The first three numbers are the direction to the nearest 10° and the next two the speed in knots. G means Gusts. 000000KT is calm, VRB means variable, less than 3 kts.

Horizontal Visibility

The minimum is in metres, followed by one of the eight points of the compass if there is a difference in visibility by direction, as with 4000NE. If the minimum visibility is between 1500-5000 m in another direction, minimum and maximum values, and their directions will be given, e.g. 1400SW 6000N. 9999 means 10 km or more, while 0000 means less than 50 metres.

Runway Visual Range (RVR)

An RVR group has the prefix R followed by the runway designator, then an oblique stroke followed by the touch-down RVR in metres. If RVR is assessed simultaneously on two or more runways, it will be repeated; parallel runways are distinguished by L, C or R, for Left, Central or Right parallel respectively, e.g. R24L/1100 R24R/1150. When the RVR is more than 1500m or the maximum that can be assessed, the group will be preceded by P, followed by the lesser value, e.g. R24/P1500. When less than the minimum, the RVR will be reported as M followed by the minimum value, e.g. R24/M0050.

Present Weather

Any precipitation. A minus (-) means light, no sign is moderate, and + means heavy. It is described in 7 ways, such as SH for showers, DR (drifting), FZ (freezing), MI for shallow, BL (blowing) and BC for batches. See under TAFs for others.

Cloud

Up to 4 cloud groups may be included, in ascending order of bases. A group has 3 letters for the amount (FEW = 1 to 2 oktas, SCT, or scattered = 3 to 4 oktas; BKN, or broken, = 5 to 7 oktas, and OVC, or overcast = 8 oktas) and 3 for the height of the cloud base in hundreds of feet *above ground level*.

Apart from significant convective clouds (CB) cloud types are ignored. Cloud layers or masses are reported so the first group represents the lowest individual layer; the second is the next individual layer of more than 2 oktas; the third is the next higher layer of more than 4 oktas, and the additional group, if any, represents significant convective cloud, if not already reported, e.g.:

```
SCT010 SCT015 SCT018CB
BKN025
```

CAVOK and SKC

CAVOK will replace visibility, RVR, weather and cloud groups when visibility is 10 km or more, there is no cloud below 5000 feet or below the highest MSA, whichever is the greater, and no cumulo-nimbus; and there is no precipitation, thunderstorm, shallow fog or low, drifting snow. Otherwise, the cloud group is replaced by 'SKC' (sky clear) if there is no cloud to report.

Air Temperature and Dewpoint

Shown in degrees Celsius, separated by an oblique stroke. A negative value is indicated by an 'M' in front of the appropriate digits, e.g. 10/03 or '01/M01'

Pressure Setting

QNH is rounded down to the next whole millibar and reported as a 4-figure group preceded by 'Q'. If less than 1000 Mb, the first digit will be '0', e.g. 'Q0993'.

Recent Weather

Significant weather seen since the previous observation, but not currently relevant, will be reported with the standard present weather code preceded by the indicator 'RE', e.g. 'RETS'.

Windshear

Included if windshear is reported in the lowest 1600 feet, beginning with 'WS': 'WS TKOF RWY20', 'WS LDG RWY20'.

Runway State

For snow or other runway contamination, an 8-figure group may be added at the end of the METAR.

Trend

For when significant changes are forecast during the next 2 hours. The codes 'BECMG' (becoming) or 'TEMPO' (temporarily) may be followed by a time group (in hours and minutes UTC) preceded by one of 'FM' (from), 'TL' (until) or 'AT' (at). These are followed by the expected change using the standard codes, e.g. 'BECMG FM 1100 250/35G50KT' or 'TEMPO FM 0630 TL0830 3000 SHRA'. Where no such significant changes are expected, the trend group will be replaced by the word 'NOSIG'.

DENEB

Fog dispersal is in progress.

Area Forecast

Covers several hundred square miles. Cloud bases are *above sea level*.

SIGMETs

Warning of serious weather.

PIREPs/AIREPs

Reports by pilots, commencing with UA. UUA is urgent.

AIRMET

A telephone service for people without access to charts or other information, with information in plain language for particular areas. See the AIP MET section for contact numbers.

Airmets are valid for 8 hours, and are issued 4 times a day, with an outlook period of 6 hours:

```
0500-1300 UTC, to 1900
1100-1900 UTC, to 0100
1700-0100 UTC, to 0700
2300-0700 UTC, to 1300
```

Charts

Weather information is issued in many ways, including the charts mentioned below. Those showing *expected* patterns are *prognosis* charts, but they won't be anything like what you see when you get there.

Form 214

A *spot wind* chart that shows you what winds you can expect at selected heights. You will find boxes placed on certain intersections of latitude and longitude, with details of wind direction, speed and temperature between 1000-24000':

5730N 0230W			
24	070	15	-40
18	060	15	-28
10	050	25	-10
05	050	15	-02
02	050	15	+06
01	030	15	+09

The location of the box is shown at the top. The column on the left shows the altitude in thousands of feet, the second the wind direction in °T, then the wind speed in knots, followed by the temperature in °C.

To get the answers to any exam questions, you have to interpolate between the surrounding boxes and the figures given in the box itself, as you can guarantee that the level given will not coincide with any in it!

For example, at 14,000 feet in the table above, the wind velocity will be 055/20.

Form 215

This is the *Significant Weather Chart*, or, in other words, a low level forecast, graphically represented. Significant weather will be shown in scalloped lines, located in *Zones* that are identified with a number in a circle. Inside the scalloped area will also be boxes with the freezing level in thousands of feet amsl.

In most cases above, the freezing level is at 6000 feet. There is an occluded front through the area, moving South easterly at 15 kts.

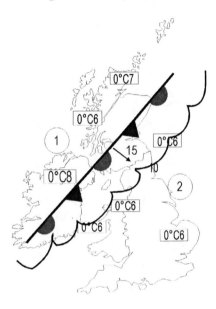

The Station Circle
Weathermen use a *station circle* (or *model*) to describe the weather where their observations are made.

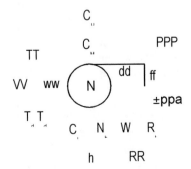

Here is a decode:

N	Total cloud
dd	True wind direction
ff	Wind speed in knots
VV	Visibility in miles
ww	Present weather
W	Past weather
PPP	Pressure in mb
TT	Air temperature
N_h	Sky covered by low or middle cloud
C_L	Low cloud
h	Cloudbase
C_M	Middle cloud
C_H	High cloud
T_dT_d	Dewpoint
a	Barograph trend
pp	Pressure change in last 3 hours
RR	Precipitation
R_t	Time precipitation began or ended

In the example below, the dark circle below means the sky is covered with cloud (it would have white and black quadrants otherwise):

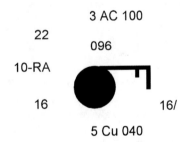

The temperature is 22°C, and the dewpoint 16°C. The wind is from the East at 15 kts – if the wind is calm, there will be a second circle around the first. The visibility is 10 sm in rain. The middle cloud is 3/10 Alto-Cu at 10,000 feet. Low cloud is 5/10 Cumulus at 4,000. The sea level pressure is 1009.6 and the trend over the last 3 hours is a steady increase of 1.6 mb.

Front Passage

You can tell the position of fronts between stations just by looking at the station circles. These items will change as a front goes by:

- *Wind* veers (marked clockwise change in arrow direction).

- *Pressure* drops as fronts approach, steadying after a warm front and rising after a cold front.

- *Temperature* changes according to the type of front.

- *Weather* will start with moderate continuous rain ahead of the warm front to drizzle in the warm sector, followed by heavier intermittent or continuous precipitation at the cold front, then nil (or showers) afterwards.

- *Clouds* follow the pattern in the diagram on the previous page.

- *Visibility* improves markedly behind the cold front.

So, if somebody asked you to give reasons why you would suspect a front in any position on a map, you would check the above elements.

Some Questions

1. Complete the following table:

PA	Deviation	OAT
FL 100		-5°C
FL 125	ISA -3°C	
	ISA +3°C	-38°C
FL 310		-52°C
FL 80	ISA +15°C	
	ISA -8°C	-21°C

2. At FL 100, what is your clearance over high ground of 5880 feet amsl, assuming 1 mb = 30 feet and an altimeter setting of 989 mb?

3. You are flying at 2500 feet near an airfield on an altimeter setting of 29.38". What is your separation from an aircraft flying overhead at FL 35?

4. If you are flying in the Northern Hemisphere, how is your true altitude changing when:

- flying over land at 2000 feet, into a headwind.

- flying over the sea at 500 feet with a tailwind.

5. If the wind at 1000 feet at an aerodrome is 360° and 15 kts, what is it likely to be at the surface?

6. If you were heading for a coastal aerodrome with a sea breeze blowing from the South, and the ETA was in the late afternoon, would you expect to land on runway 36, 18 or 21?

7. Flying towards a warm front above the freezing level, you encounter rain. What sort of icing are you most likely to get?

8. Your destination has fog in the early morning, with hardly any wind. If the wind increases to 10 kts, what can you expect when you arrive?

9. A TAF time group of 0220 means what?

10. A VOLMET report for 0500 UTC in the Autumn gives a surface wind of 5 kts, temperature of 9°C, dewpoint of 8°C and 1500m visibility, with no cloud reported. If you plan to arrive at 0600, what weather can you expect?

11. What do these parts of a TAF mean?

```
0615 15030G401200 BR
SCT008
```

12. And these?

```
TEMPO 1420 8000 SHRA
PROB30 TEMPO 1415 5000
```

13. On a calm, clear evening, the METAR for the destination contains the figures 04/03. What can you expect on arrival?

14. What is QFF?

15. Define Density.

16. Define Dewpoint.

17. What is the SALR?

18. What is a gust? A squall?

19. At what height is the geostrophic wind?

20. What is an isallobar?

21. What is an anabatic wind?

22. What might you expect from a thunderstorm?

23. What might you expect from icing?

Some Answers

1. ISA 0°C, -13°C, FL 280, ISA -5°C, +14°C, FL 140.

2. The difference between the altimeter setting and 1013.2 mb (29.92") is 24 mb, a difference of 720 feet, so the PA of the high ground is 6600 feet. Subtract this from 10,000 feet for 3400 feet clearance.

3. 29.38 from 29. 92 is .54, or 540 feet, so the PA of your aircraft is 3040 feet. The separation is therefore 460 feet.

4. The answers in order:

- At 2000 feet, wind is unaffected by surface friction. With no drift, you are parallel to the isobars and true altitude is constant.

- True altitude is decreasing, because the 500-foot wind is backed with reference to the wind at 2000 feet, and you are crossing the isobars from high to low pressure because of the drift you have to apply.

5. Wind speed will reduce by about 20%, so the coriolis effect will automatically reduce by about 20°. Thus, 360-20 is 340°, and 20% of 15 is 3, so the speed will be 12 kts.

6. At first sight, it would be 18, but in late afternoon, the Coriolis effect could well make it 21.

7. The airframe is cold, as you are below freezing. You will therefore most likely get clear ice.

8. Thin low stratus cloud.

9. It is valid between 2 in the morning and 8 at night.

10. As the wind is light, and the temperature is very close to the dewpoint, you are very likely to see radiation fog, because there is no cloud and it is very near dawn. The visibility is already poor, and will get worse, before possibly clearing by mid-morning after solar heating. Note that the minimum RVR for a visual approach is 800m.

11. Valid between 6 in the morning and 3 in the afternoon (UTC, of course), the wind is from 150° at 30 kts, gusting 40, visibility 1200 metres in mist. 3-4 oktas of cloud at 800'.

12. Temporarily between 1400 and 2000, 8000m visibility in showers and rain. 30% probability of a temporary reduction in vis to 5000m between 1400 and 1500.

13. The temperature is 4°C and the dewpoint 3°C, meaning that, if the temperature falls, as it is likely to if it is calm and clear, moisture will condense out and form mist or fog.

14. The QFE reduced to mean sea level pressure using the ambient temperature lapse rate. It gives a more accurate estimation of sea level pressure and is used when producing surface weather charts.

15. Mass per unit volume.

16. The temperature to which air must be cooled, without change of pressure, for the air to just become saturated. Any further cooling results in condensation.

17. The *Saturated Adiabatic Lapse Rate* is the rate at which a parcel of saturated air changes temperature with height.

18. Temporary increases in prevailing windspeed. A squall is normally associated with thunderstorms and lasts longer.

19. Around 2000 feet.

20. A line joining points of equal rate of change of pressure.

21. The reverse of a katabatic wind, flowing uphill (at a slower pace).

22. Hail, icing, lightning, static electricity, turbulence and windshear.

23. More weight, less lift, more drag, stuck controls, less visibility, radio interference and blockages.

Electricity & Radio

Radio depends on the movement of electric and magnetic waves, which depend on the movement of electricity, which ultimately depends on the activities of electrons inside an atom.

Atomic Theory

Electrons are negatively charged particles spinning rapidly round a *nucleus* of positive- and neutrally charged ones, called *protons* and *neutrons*, respectively, as shown below (the neutrons keep the protons together, since particles of a like charge are repelled):

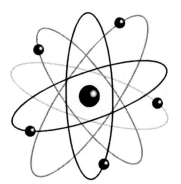

None of them are physical in nature, but are actually electromagnetic charges, or tiny whirlwinds of electromagnetic force. A collection of atoms is called a *molecule*, which is the smallest part of any object that retains the identity of it.

Put very simply, if you line up a series of atoms (as in an electrical cable) and add an electron to the first one, it will repel those already there until one is pushed out, which joins the next atom, and so on down the line until an electron falls off the last one, giving you an electric current. When the electron is pushed in at first, there is a difference in potential between that end of the cable and the other end, creating a *potential difference* (another name for voltage). In other words, the flow of electricity is like that of wind moving from high to low pressure.

Some atoms don't have much of a hold on their electrons, and allow them to move easily – the materials made up of these are called *conductors* (copper is a good example – gold is only used because it doesn't tarnish in a hurry, and cause bad connections. A gas can also conduct electricity). Those that keep a tight hold and therefore allow no movement are found in *insulators*, which are used to keep conductors from touching each other, otherwise electricity would flow where you don't want it – if electricity takes a short cut (known as a *short circuit*) it generates massive amounts of heat, with the obvious consequences. Good examples of insulators would be glass, or the plastic coating round a cable.

Somewhere between a conductor and an insulator is a *semiconductor* which is created by adding a certain amount of impurity to a material normally considered to be an insulator. Electricity will then flow only under certain circumstances, such as the influence of an electromagnetic field, or the polarity of the source of a current (this is the basis of the transistor).

There must be an equal number of electrons to protons, which is why an extra electron (or a hole caused by one leaving), is balanced immediately. An atom with one extra is *negatively charged*, and with one missing is *positively charged*. This process is called *ionisation*, because an atom wrongly charged is called an *ion*, which we will come across latter when we look at the *ionosphere*, or what is now called the *mesosphere*. Some components, like transistors, depend on the movement of

electrons or holes (missing electrons) one way or the other.

Electrons spinning round an atom occupy *energy levels*, or *shells*, rather like the orbits of the planets around the Sun. The first shell can hold up to 2 electrons, and the second up to 8, but it's always the outer shell, which contains *valence electrons* that is important. Such atoms can be dislodged easily by applying stress in the form of heat or a magnetic field, which is how an electrical current is produced.

Electricity

There are three types of electricity:

- That which stays right where it is, called *static electricity*.

- That which goes in one direction only, usually at one speed, called *Direct Current*.

- That which flip-flops back and forth to form a wave pattern, or *Alternating Current*.

The essential point is *movement*, since nothing much happens when everything is still, but it is the last one, AC, with which we will mainly be concerned.

Batteries

Certain chemicals when combined with metals can cause electrons to flow as direct current, until all the electrons disappear from the metal, causing it to eventually get eaten away – since the atoms comprising it lose electrons, they cease to be the same atoms and therefore cease to exist in their former state – if you could contrive to put the electrons back, you would regain your metal plate, and recharge the battery.

Actually, a battery is a collection of *cells*, which typically have a charge of about 2 volts each when it comes to aircraft, hence the need to combine them in order to do anything useful. Knowing how to do this is handy when you are out in the field with a discharged battery and you are trying to connect two car batteries together to start your aircraft.

The polarities are *positive* and *negative*, marked as *plus* (+) or *minus* (-), or red and black. If you join batteries in *series*, that is, one after the other, with the positive of one connected to the negative of the next:

you will get a voltage which is the *sum* of the cells, but with the amperage rating of one cell. If you join them in *parallel*, with the positive and negative terminals connected to each other:

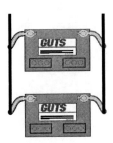

you would get the voltage of one cell, but the amperage of all of them (see above).

Since a typical aircraft runs on a 24-volt system, you would therefore connect two car batteries *in series*. Be aware, though, that the terminals are different sizes to stop them being confused with each other, so you need to carry an adapter in your navbag to connect them up in the middle (try Halfords). Be wary of jumper cables, as they may open up and spark when a load is applied. Ensure that batteries have a load on them before completing a circuit.

A battery cell is made up of *electrodes* surrounded by *electrolyte*. Different materials are better or worse at this job, so you might get more or less voltage out of one type of battery or another, but the most common is zinc-carbon (there are also alkaline, mercury, NiCad, Lead-Acid, Lithium, etc.). Electrons flow from the *negative* (-) electrode, through whatever circuit the battery is connected to, back to the *positive* (+).

Primary cells are not rechargeable, but secondary ones are.

A flat battery has maximum internal resistance, which will generate lots of heat when the aircraft alternator or generator attempts to charge it (on a bench, only a *very* small current is used). It is therefore not a good idea to continue flight if your battery gets discharged – in any case, *it should be replaced before the next flight*.

The battery itself will be rated in terms of *Amp/Hours*, meaning that it's supposed to provide a certain number of amps for a certain number of hours when fully charged, though it is never wise to rely on *any* battery for more than about 20 minutes (officially, they should last for at least 30 minutes). To get an idea of your aircraft's capabilities, add up the number of devices that use power (check the circuit breakers) and divide them into the amp/hour rating. So, if your devices collectively use 45 amps (see below), and your battery supplies 45 amps/hour, you should be able to get an hour out of it. You could also use half the devices and get two hours. When faced with such an emergency, it is usual to use the navaids, for example, to get a position fix, then turn them off until you start feeling a little lost, then turn them on again until you are once more certain of your position. The same with radios. This will get a little extra time out of your battery.

There are two types of battery used in aircraft, *lead acid*, as found in cars, and *NiCad*, as found in portable computers. People who use both will already understand the difference, but just in case, the lead acid's output tends to fall off with discharge, whereas a NiCad can pump out power at a constant rate until it can do no more, as well as recovering more quickly. The trouble is that NiCads have short memories, in that if you keep charging them up when they have only discharged a little way, they will begin to think they have a lesser power rating, so to stop them causing hot starts they need regular *deep cycling* to keep them awake. So, although it's good practice to start a helicopter, for example, from a battery cart, to preserve the ship's battery for better

reliability in remote places, occasionally a battery start is good for the system.

Another problem with NiCads is that they can spontaneously combust when too much current is drawn and then replaced (actually called *Thermal Runaway*). This is why some helicopters have a *Battery Temp* caution light on the warning panel which means you must land *immediately*, before the battery catches fire and takes other stuff with it, if it doesn't actually burn its way through the airframe and fall out.

If you spill any electrolyte from a NiCad, you can neutralise it with dilute boric acid.

The *Battery Master Switch* controls the power to all circuits, and there will be other switches to control smaller groups of equipment, such as the *Avionics Master Switch*, for the radios and navaids. The Battery switch may well be in two parts, one for the battery itself, and the other for the alternator circuit. Circuits will be protected by fuses or circuit breakers, which *should only be reset once*, since there is a reason for them blowing the first place.

A fuse is a deliberately weak part of a circuit that is designed to fail if a problem should happen, thus protecting the rest of the circuit and saving the trouble of replacing wiring in odd places.

A more modern replacement is the *circuit breaker*, which is a button that pops out when a fuse would otherwise break. A *trip-free* circuit breaker is one that will trip even if it is held in.

The voltage regulator is there to stop the battery being overcharged or the system being overloaded by the generator or alternator. There will be a warning light in the cockpit to indicate that this is happening, and that you are getting battery power only. The *over voltage sensor* may be reset with the Master Switch. It may be tested by turning the ALT half of the battery switch off for a moment.

When starting, the starter switch will activate a *solenoid*, which is just a bigger switch that can handle more current, to actually turn the engine. Since the current is large (60 amps), there is no fuse protection, which is why there is a starter light.

Amps

The flow of electrons in a conductor (i.e. the *current*) is expressed in terms of *amperes*, or *amps*, which are defined as the movement of 1 *coulomb* per second (a coulomb is the accumulated charge of a large number of electrons, actually 6.28 x 10^{18}). In an aircraft, amps would be measured with an *ammeter*, or *loadmeter*, a useful device for checking if your battery is being charged. An ammeter needle should always be showing in the + side of the gauge, showing a positive charge.

Volts

The work done to add electrons to an atom is expressed in *volts*, which move from high to low pressure, like air does. Once a body is charged this way, it is "pressurised" (for want of a better word), and the potential energy it contains is called the *potential difference* when it refers to a difference in energy, or pressure,

between two points. You can look on volts as the equivalent of water pressure.

Another (older) name for volts is *electromagnetic force*, or *emf*. It is measured with a *voltmeter*, which you might use to check the state of your battery before starting a jet engine.

Resistance

Even a good conductor slows electrons down. The longer and thinner the wire is, the more the opposition, called *resistance*, expressed in *ohms*. 1 ohm allows 1 amp to flow when 1 volt is applied.

All this work causes heat, due to the friction of electrons moving against each other, and the more work you make electricity do, the hotter things get, which is how electric fires work. If you make it work harder, you get light as well, hence light bulbs.

When you start using AC, however, the current flows on the *outside* of the cable, increasing the resistance (many times) because the effective cross-sectional area is reduced, which is called the *skin effect*.

Magnetism

A magnet is a substance, typically a soft iron bar, that has lines of force running through and around it (the Earth is a magnet as well). All magnets have a North and a South pole, and two North poles will repel each other. North and South poles attract each other. If you therefore had a bar magnet, its *South* Pole (traditionally red) would point towards the Earth's (magnetic) North pole. This is what a compass is all about, discussed more fully under *Instruments*. The thing to remember, though, is that the South Pole is marked as North.

Magnetic reluctance is the ability of a substance to pass lines of flux within itself. Hard iron in this respect will not pass flux easily, so it has a high reluctance and is therefore not easily magnetised. *Magnetic permeability* is the opposite, characterised by soft iron, which is easily magnetised.

A magnetic field moving round a conductor will induce an electrical field in the conductor – the more the flow, the bigger the field, which follows the direction of flow. You can get the same effect by moving the conductor in the field, or by moving them both together at the same time.

This works in reverse as well – a flow of electricity in any conductor produces a magnetic field around it. If you put a coil of wire round a soft iron bar, and run electricity through it, the iron will become a magnet for as long as the current flows.

During engine operation, the battery will be recharged with either a *generator* or an *alternator*, based on DC or AC, respectively, which will put out more voltage than the battery (typically 28v in a 24v system, or 42v in a 36v system) to make sure the battery doesn't drive the generator. The alternator will charge at low RPM, but some helicopters, notably the Bell 206, use a *starter/generator* to save space, despite this advantage (the same unit is used to spin the engine on startup, and switched over when it's running to become a generator). If an alternator were used, you would need yet another item attached to the engine. Generators therefore have the disadvantage of not producing lots of electricity at low engine RPM (exam question).

Direct Current

As previously mentioned, this electricity that goes in one direction only. It can be produced in many ways, such as *friction, heat, pressure* and *photoelectricity*, but we are concerned with *magnetism* and *chemical action* (see *Batteries*, above)

The Generator

These use magnetism to create DC. A coil of wire round soft iron core (an armature) is spun in a magnetic field to cause a current to flow in them - in fact, the current starts off as AC and is converted with a *rectifier*. The lines of magnetic force either come from a permanent magnet or an electromagnet formed from the generator's own current (in which case it is called *self-excited*, which can happen in the first place through *residual magnetism*). When the generator's field is provided through the Master Switch, generator voltage will drop to zero when the switch is turned off (Master Switches are mainly there for alternators, which are not allowed to produce current without a battery in the circuit).

Because the windings are in the same direction, they are in series, and will create a voltage equal to their sum.

Alternating Current

As mentioned above, this is electricity that reverses its polarity (and direction of flow) several thousand times a second.

Again, as mentioned above, this is electricity that reverses its polarity (and direction of flow) several thousand times a second.

In an aircraft, it is typically used to power flight instruments, or to charge the battery in preference to a generator, which does not work so well at low engine RPM (exam question). AC is more efficient at transmitting energy over long distances, and smaller conducting elements are required. However, alternators cannot self-excite – they need battery power to function. The higher alternating frequencies for radio and radar (below) are produced with *oscillators*.

If you can imagine changing the connections to a battery very quickly from one terminal to the other, you would get the same effect, but the results would be very jerky, and the waves would be square. In contrast, the transitions from an alternator are very smooth and look like *sine waves*, as in the diagram below. As it has only one frequency, if you could listen to it, it would sound like a continuous tone.

The rate at which it does this is called the *frequency* which, in a typical North American home, is 60 *cycles per second*, or 60 *Hertz* (1 Hertz is equal to one cycle per second). In Europe, it is 50 cycles per second, which means that electric motors based on 60 cycles won't work properly.

A *cycle* is a complete transition from zero through a peak, down to a trough and back up to zero, so the more you can fit into a particular

time scale (the higher the frequency), the shorter the *wavelength* is:

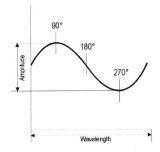

The rate at which it does this is called the *frequency* which, in a typical European home, is 50 *cycles per second*, or 50 *Hertz* (1 Hertz is equal to one cycle per second).

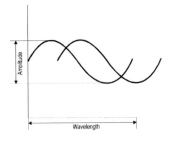

The difference between the peak (or crest) and the trough of a wave is the *amplitude* (or, loosely, volume).

One complete transition of a wave, no matter where it starts, is called a *cycle*. The number of *cycles per second* is known as the *frequency*. When two waves transmitted at the same time have their peaks and troughs coinciding, they are said to be *in phase*. When they don't coincide, they are *out of phase* by whatever angle is created when the second wave start its cycle:

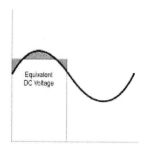

This is the basis of finding direction with the VOR, discussed later.

Because the AC waveform is not square, its peak voltage must be 1.414 times that of DC to have the same energy. The equivalent AC voltage to match a DC one is called the *root mean square*.

In the diagram above, the shaded area inside the curve over the square DC wave has the same energy as the shaded parts of the DC wave outside the curve. In your house, the peak would actually be 340 volts, as opposed to the "normal" 240.

The Alternator

AC is created in the first place with an *alternator* (or an *inverter,* which converts DC). On modern aircraft, however, the term *A C generator* is often used instead.

The other relevance of AC is that it is the basis of radio waves ....

Radio

We use radio to convey information. For example, sound waves by themselves don't travel very far, so what you do is create a carrier wave at radio frequency and piggy-back a sound wave on to it, decoding the two at the receiving end–a radio is a device for splitting up the two waves and amplifying the result, since the signal when it hits the aerial is quite weak. Inside the box, therefore, are several separate devices that work together, needing different amounts of electricity, including a *resonant circuit* to ensure that you pick up only one signal out of the many available.

The information to be sent *modulates* (i.e. varies) the carrier wave through the *amplitude* or the *frequency*. The former is typically used in aviation, and the latter by FM music stations (i.e. *frequency modulation*), which is less subject to interference. Many customers, such as forestry, also use it for communications. Anyhow, a modulator's job is to combine signals from the radio and audio amplifiers.

The simplest method of transmitting information is to turn a signal on and off in a recognisable code, as used by older NDBs which break the signal in a pattern matching the Morse Code ID of the station, called *wireless telegraphy,* or *continuous wave (CW)*. This is known as an A1 transmission, whereas a carrier wave by itself would be known as A0.

Note: Although Marconi transmitted the first CW signal, a Canadian, Reginald Fessenden, transmitted the first *voice* signal from Massachusetts to ships along the Eastern Seaboard. Mind you, Nikola Tesla was way ahead of them both.

Otherwise, you can adjust the *amplitude* (or volume, if you like), the *frequency,* or the *phase,* if the frequency of both waves is the same. Simple AM is known as A2, and complex AM (i.e. voice) is called A3. Jumping a bit, the VOR, mentioned below, is A9W, because its carrier wave varies with amplitude and the frequency at the same time.

Wavebands

The range of electromagnetic waves is quite large, but radio waves only occupy a small part of it, actually between about 3 KHz to 3,000 GHz. This area is split up by International agreement between the people who wish to use it, and consists of frequency ranges that share similar characteristics:

- **3-30 KHz**. VLF – Very Low Frequencies, with *very long waves*. Used by Omega.

- **30-300 KHz**. LF – Low Frequencies, with *long waves*. Used by Decca, NDBs, Loran.

- **300-3,000 KHz**. MF – Medium Frequencies, with *medium waves*, used by most AM stations, and NDBs, with static problems.

- **3-30 MHz**. HF – High Frequencies, with *short waves*, used for long-range SSB communications between aircraft and ground stations, with static problems.

- **30-300 MHz**. VHF – Very High Frequencies, used with amplitude modulation for voice comms, etc. Relatively static-free, needs line of sight. Used also for VOR, ILS Localiser.

- **300-3,000 MHz**. UHF – Ultra High Frequencies. DME, SSR, ILS Glidepath, GPS.

- **3-30 GHz**. SHF – Super High Frequencies, with *centimetre waves*, as used in radar. Also known as *microwave frequencies*, so good for MLS, Radar, Doppler and radio altimeters.

Propagation

An invisible connection between two points is called a *field* – since radio depends on the interplay of electricity and magnetism, there is an electromagnetic field joining your radio with whatever is transmitting.

A change of one type of field causes a change in another, so if you vary an electric field, it will induce changes in a magnetic field and *vice versa*, which is how an aerial is used to transmit – flip-flop movement of electricity up and down creates a magnetic field around it, and the movement of the magnetic field creates an electric field, which creates another magnetic field, and so it goes on until the power fades in an inverse square relationship, meaning that a signal 2 nm from its source will have a quarter of the strength of one only 1 nm away.

The result is an electromagnetic wave with of one of each type at right angles to each other (a wave's *polarisation* is noted with reference to the electrical field, so a *vertically polarised* wave has a vertical electric field, which will come from a vertical aerial. For efficiency, the receiver must have the same orientation).

The trick is to flip-flop the electricity so fast along the antenna that it effectively falls off the end and keeps on going, which doesn't happen below a certain frequency (the frequency of the field is the same as the AC along the antenna).

A *space wave* may leave the antenna at an upward angle, or be bounced off the ground. If contained within the troposphere, it will also be known as a *tropospheric wave*, or *Direct Wave*. Otherwise it will be a *sky wave* when headed for the *ionosphere*, where it might be bounced downwards again, if the angle is right, and reach further distances (on HF). The ionosphere is a region where the Sun's UV rays dislodge electrons from the gas molecules, making them positively ionised. This happens mostly during the daytime and is at its minimum just before sunrise.

Anyhow, any wave that hits the ionosphere is bent, as the side of the wave that hits a layer first starts to speed up, which has the bending effect. Eventually, if the angle is increased, the bending will be enough to bounce the wave back to Earth (we won't get into Moon bouncing here!). The angle when this first happens is called the *critical angle*.

A *ground* or *surface wave*, in contrast, may go directly, or curve to follow the Earth's surface, depending on the frequency. Friction with the ground and the widening circumference of the wave will eventually weaken its power, though. When a wave leaves an antenna, the ground wave will be detected until it fades, or *attenuates*. Between that point, and where the first sky wave comes from the ionosphere, is an area where nothing is heard, called a *skip zone*, or *dead space*.

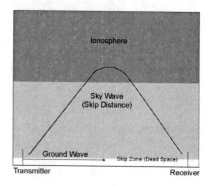

Surface and atmospheric attenuation *increase* with frequency, while ionospheric attenuation *decreases*. Ground range *increases* if critical angle, frequency, dead space and skip distance *decrease*, and *vice versa*.

The *skip distance* is the Earth distance taken by a signal after each *refraction*, or the distance covered by the first sky wave. The *maximum useable frequency* exists where skip distance is the same as that between the transmitter and receiver. The *optimum useable frequency* is about 85% of that, to allow for variations in the height and thickness of the ionosphere (see *VOR* for the formula).

The ionosphere moves all the time, affecting it considerably, which is why the ADF suffers from what is called *night effect* just after sunset and before sunrise when the needle swings erratically (on the other hand, during night is when you will receive distant stations best). Refraction can occur in many ways, from the ionosphere (see above), the coast (a wave crossing at anything other than a right angle will be bent) through to the atmosphere – under certain inversion conditions, a phenomenon called *ducting* occurs, which enables waves to travel unusually large distances (the author has certainly seen British TV programs in Germany with this effect).

Fading is due to *multi-path propagation*, amongst other things, where signals may be received from many sources and be out of phase with each other at the aerial. Sometimes, under such circumstances, waves will cancel each other out.

Ground waves are associated with LF/MF waves, *sky waves* with HF and *direct waves* with VHF frequencies and above. In fact, the latter are also called *line-of-sight*, meaning that anything in the way, like hills or buildings, will have a detrimental effect (they will not bounce like HF). You will get best reception if the transmitter and receiver are in sight of each other, but, in practice, you can expect a little more than that, actually to just beyond the horizon, due to effects like refraction within the troposphere (see *VOR*, below). The actual figure is greater by a factor of around 4/3.

Air-ground transmissions are limited to 25 nm in the UK, up to 4,000 feet for tower frequencies and 10,000 feet for approach.

HF frequencies need to be higher during the day or when you are at greater range from the station. At night, you can use lower frequencies, generally about half (that is, use *Double During Day*).

VOR

This stands for *Very High Frequency Omnidirectional Range*, so is based on VHF, using the phase difference between two signals to signify your direction from the transmitting station. The frequency range is between 108-112 MHz on even decimals, plus 50 KHz (to prevent confusion with the ILS), and 112-118 on odd and even, plus 50 Khz. VORs are identified on maps with a compass rose around the station aligned with Magnetic North. It is *not* sensitive to heading, as is the ADF (below) - it shows *track*.

The *Station Identifier* is transmitted in Morse Code every 15 seconds, and you must confirm the frequency and ID before using a VOR for navigation. If there is no ID, but behaviour is otherwise normal, the system is on maintenance.

The transmitter sends out a reference signal in all directions, frequency modulated at 30 Hz, which is received by all stations at the same phase, if they are the same distance away. However, it is not transmitted at the same *strength* all the way round – the amplitude is also varied to produce a polar diagram called a *limacon*, which is similar in shape to the *cardioid* used by the ADF (below), but without an absolute null point. It rotates 30 times per second. The phase of this signal in the aircraft depends on the bearing from the station (which is probably why it's called a *variphase signal*). Both signals are in phase at 0°, or North, 90° out when East, and so on. For each degree moved, the signal changes phase for the same amount, in both frequency and amplitude, which is how your direction is determined. Because the signal is frequency and amplitude modulated, it is classed as an A9W signal. Just to complicate matters, Doppler VOR has its modulations the other way round.

Overhead the beacon, you will be in a *cone of confusion* that exists with any antenna – this is an area where no signal is received, so the TO/FROM flags disappear and the alarm flag comes up (in the case of the VOR, the cone is 100° across). During this *station passage*, you should ignore the signal. There will also be ambiguities abeam the beacon - at a point 90° either side of the selected radial there is a *zone of ambiguity* where the flag will not show at all, and the indications should therefore not be relied upon.

Inside your aircraft will be a large black box somewhere in the back, with a remote indicator in the cockpit:

Once you select a radial by turning the *Omni Bearing Selector* (the small knob just under the dial), the *Course Deviation Indicator* needle will be in the centre, or either side of the centreline, up to 10° away from the radial, so each dot left or right represents 2°, if there are 4 dots on your display (2 ½° if there are 3). When in the middle, you will be on the *radial*, which traditionally is *from* the station when on the same side, shown by *To/From Flag*, which, on later instruments, will be a small white triangle pointing in the relevant direction. If the indicator

shows *To*, you are on the *reciprocal*, or on the other side (in the above example, the radial is N, or 360°, because the *To* flag is showing).

This is a common trap in exam questions – if you are tracking inbound on a radial, remember to set the *reciprocal* at the top of the display, since radials go *from* the station.

All you have to do then is watch the needle – if you are going away from a station on a radial, and the needle is pointing left, then you fly left until it centres:

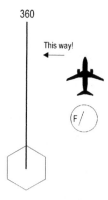

If you are going to the station, then you fly right. The thing to remember is that *the needle always points to where the radial is*, which is *nothing to do with the heading of the aircraft* (remember this for exams). All you do is follow the needle (when coming down the

ILS, you follow the cross formed by the localiser and glideslope needles). In short, the radial is where the needle is, and you do not necessarily turn that way to get to it - sometimes, having the needle on the left means turn right! Only if your heading is the same direction as the OBS will it be on the correct side.

For any radial, there are boundaries formed by the CDI and the To/From indicator, forming quadrants around the station (that is, there are four distinct areas). You will be in one of them.

You can therefore take the indications *from* two VOR stations, draw the *lines of position* (i.e. *bearings*) from the compass roses and the intersection point is your position:

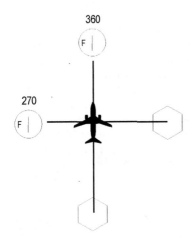

Remember to ensure that the CDI is centred in both cases and the FROM flags are showing.

When tracking along an air route, tune and identify the station you are going from, track the selected radial until near the mid-point, then tune and identify the next station. The To/From flag should change over.

If you have to use another VOR to provide a fix as a reporting point along the air route, select the required radial, and when the needle is centred you are over the fix:

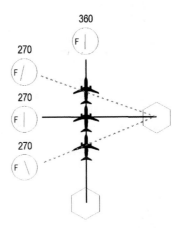

If you have to check whether you have passed it or not, in this case, the needle will be to the left of the station is on the left, and *vice versa* if you are not there yet, assuming your heading is the same as the OBS. Otherwise, the needle will point the opposite way if you have already gone past (oops!).

To intercept an inbound radial, tune and identify the VOR station, select the reciprocal of the desired radial, by turning the OBS until you get a TO reading. Fly to whichever side the needle is displaced, turning the shortest way to a heading 90° away from it, until the needle starts to move, at which point reduce the intercept angle to 45°:

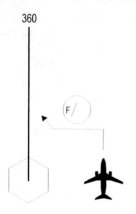

360

As the needle centres, reduce the intercept angle again and maintain the track with suitable adjustments for drift.

Do the same outbound, except look for a FROM reading.

A good rule (inbound and outbound) is to subtract the intercept angle if the needle goes left, and add if it goes right to find the heading to steer. For example, 280°-90°=190°.

To bracket for drift, turn onto a zero wind heading and see what the drift actually is. Get back on track, make a large correction the opposite way and see what happens. Get back on track and half the original correction. Keep doing this until the correct heading is found.

Testing
Some airfields have low power test equipment (2 watts) transmitting on 114.8 (usually), identified with the ATIS, so have a pen ready to save you writing it down again later (the ID may just be a series of dots). The system is intended for ground use – although it can be used when airborne, there will be certified

airborne check points, but you could always get to a position on a known radial and check the readings. As you move the OBS, you can expect the usual indications relating to the bearing selected (which is why two transmitters are used, to save you moving the aircraft to the radials). With the needle centred, the instrument should read 000° FROM or 180° TO at any point within the airport, with an accuracy of ± 4° (± 6° when airborne).

In fact, *transmitter error* (or FM/AM synchronisation, at least) should be within ±1° - the system should shut down automatically if it gets outside that. *Phase comparison error* should not be more than ±3°, and *station errors* should be within ±1°. The nominal accuracy is ±5°.

Problems
Although the VOR is less subject to static and interference than an NDB, and it is much more accurate, the transmissions depend on line of sight, and there is a zone of ambiguity at 90° to a radial, mentioned above. In addition, certain propeller or rotor RPM settings can cause fluctuations up to ±6° (change the RPM slightly before saying the instrument is U/S).

Time to Station
You often need to know the time it will take to get to a station, which is simply found by turning 90° from the inbound radial and noting the *seconds* taken to go through a number of them. To get the time in *minutes*, divide the time just noted by the number of radials (degrees) gone through. All you need do then is use

the groundspeed (or TAS in emergency) to find your distance.

For time to station, the formula is:

```
Time (mins) = Mins x 60
              Degrees
```

On the whizzwheel, set the minutes on the outer scale, and the degrees on the inner one. Read the answer on the outer scale opposite the 60 arrow.

For the distance, try:

```
Distance = Mins x GS
           Degrees
```

ADF/NDB

An *Automatic Direction Finder* (ADF), also known as a *radio compass*, is a device in an aircraft that picks up signals broadcast on the Medium wave band by *Non Directional Beacons* (NDBs), so called because they radiate in all directions.

Transmissions are not dependent on line of sight, so the system is good for long distance travel, although it does have a few problems, mentioned below. It is possible to get 1,000 nm range over sea and 300 nm over land if the power is high enough, but since better systems have come along, NDBs are now used as enroute navaids on airways, homing beacons for instrument approaches and markers for the *Instrument Landing System* (ILS), with a typical range of about 35 nm.

NDBs should be accurate to within ± 5° by day. If there is no ID, but the system otherwise appears to behave normally, the NDB is undergoing calibration or maintenance.

The primary function of the ADF receiver is to determine the bearing of an incoming NDB signal, which is vertically polarised. To do this, it uses a *loop aerial*. When the loop is square to the beacon, the signal reaches both sides of the loop at the same time and there is no signal detected. When the loop is sideways-on, however, the signal reaches one part of the loop first:

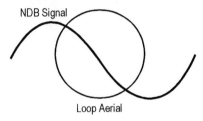

Loop Aerial

and the second part will be out of phase, so a current will be generated, which drives an electric motor to continually seek the null position. It is phase sensitive, so it can always turn the shortest way. Various stages of magnification inside the receiver help this along, but that need not concern us here. The point is that the detected signal is not actually used to determine the bearing, but the null signal point, since the current flow is slow to build up and break down, and is a bit on the woolly side. The null point is much sharper and easier to find.

Because the current flows in the opposite direction depending on the position of the loop antenna, you also need some way of determining which end is what, otherwise you could be 180° out. A single vertical aerial called a *sense antenna* helps here – both signals are combined algebraically and the magnitude and polarity of the sense aerial arranged to be identical to the loop. The result is a polar diagram called a *cardioid*, which has only one null point:

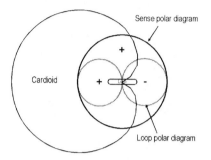

On one side of the loop the polar diagrams are positive and combine, but on the other, one is positive and the other negative. They cancel each other out, hence the null point on one side.

The modern (and more stylish) equivalent is a small housing with two coils at right angles to each other, wound on ferrite cores. They are connected to the *stator coils* of a *goniometer* which points the needle.

Limitations of the system include:

- *static.*

- *station overlap*, where NDBs have the same frequency. Because this is more pronounced at night, it can easily be confused with *night effect*, below (*promulgated ranges* are not valid at night for this reason).

- *night effect*, where the needle swings erratically, at its strongest just after sunset and before sunrise. The loop is designed to receive surface waves – any others resulting

from sky waves will be out of phase and confuse the system entirely. If the ionosphere is not parallel with the Earth's surface, they will also arrive from a different direction. Check for an unsteady needle and a fading audio signal. Promulgated ranges are not valid at night.

- *quadrantal error*, variations caused by the aircraft itself, in the same way as it might affect a compass. The signal is reradiated by the airframe and the receiver gets an additional (much weaker) signal to contend with. The greatest error lies at 45° to the fore and aft axis, hence *quadrantal*. Modern systems have corrector boxes for this.

- *mountain effect*, or variations caused by reflections from high ground, where two signals might be received at the same time from different paths.

- *coastal refraction*, from radio waves in transit from land to sea, or parallel to the coast, because they travel slightly faster over water. It is most noticeable at *less than* 30° to the coastline (exam question), and at lower frequencies. The effect is to make the aircraft appear closer to the shore.

The most common error, though, is failing to recognise *station passage* – if you are directly over the beacon, it will swing erratically and be confused with one of the above (exam question), or failure of the instrument where the needle just rotates to the right all the time. This is due to the same cone of confusion that exists with the VOR (above).

The ADF is normally tuned with the function switch in the ANT position (it stands for *antenna*). This removes the needle from the loop and saves wear and tear as it tries to point at every station you tune through – here, the sense antenna is used by itself to obtain the ID. Once there, return the switch to the ADF position (or COMP, on some sets). As always, check – in this case, ensure that the needle points vaguely where you expect it to. The TEST button spins the needle 90° away from its tuned position, and return, to indicate a good signal.

BFO means *Beat Frequency Oscillator*. The BFO switch also uses the sense aerial by itself to detect the modulated Morse identifier. Hearing this by itself helps you tell if there is any fading (night effect) or noise (thunderstorms, interference). The tone you hear when this switch is activated is actually put there by the ADF receiver, since a carrier wave by itself cannot be heard.

Aside from continuously listening to the ID, the only way of knowing there is a problem is seeing the needle rotate to the right if the signal is not received.

The *fixed card display* (goniometer) consists of a compass rose with 0° representing the nose of the aircraft at the top of the instrument, and a needle that points to where the signal is coming from (including thunderstorms if they are stronger than what you are tuned into). Thus, if a station is ahead, the needle will point to 0°, or 180° if it is behind. However, if you made no allowance

for wind, and just pointed the nose of the aircraft at the station, you would actually follow a curved path towards it. Allowing for drift lets you keep a straight track, which is needed to keep on the airway (see *Tracking*, below).

Unfortunately, working with fixed cards involves a little maths. First of all, though, here are a few definitions:

- *Magnetic Heading* – the angle between the aircraft's longitudinal axis and magnetic North.

- *Relative Bearing* - the angle between the aircraft's longitudinal axis and the NDB, which is what you read directly from a fixed card ADF.

- *Magnetic Track* or *Bearing* – the angle between aircraft position and the NDB, either to or from.

Take note of this formula (you will need it in the exam):

$$MH + RB = BTS$$

In other words, the magnetic heading plus the relative bearing gives you the bearing *to* the station.

Taking the above example, the formula would read:

$$324 + 46 = 010$$

Get the relative bearing like this:

$$BTS - MH = RB$$

If you split the display into two halves, based on a line between 0° and 180°, and call the right half *plus*, and the left *minus*, you can use the needle's position in either to find the track to a station. For example, if the needle is in the right half (the plus segment), add the heading to the relative bearing to get the track. If it is in the left, take it away (work the needle back from zero in this case).

Whilst turning right, the aircraft heading will *increase* while the relative bearing *decreases*, and *vice versa*. As long as you remain on the same bearing, the amount of heading change will always equal the change of ADF indication.

RMI

The *Radio Magnetic Indicator* is a combination of ADF indicator and slaved compass that replaces the fixed card with one that moves, so the top of the instrument represents the aircraft heading and the needle points to the actual bearing (or reciprocal, if you look at the other end), which saves you doing the calculations above in your head. In other words, it always displays the present heading and bearing, and does some of the work you would have to do with a fixed display. There will also be a repeater needle from the VORs giving you the same information relative to the stations they are tuned to.

In the above example, the heading is 324°, and the Bearing To Station (BTS) is 010°.

As a point of interest, the VOR needle on an RMI will always read correctly if any deviation occurs – the heading and ADF reading will be in error by the amount of deviation.

Position Fix

For a fixed card ADF, find the relative bearing to each station and add them to your heading to get the tracks to the stations. Then find the reciprocals and plot them outwards.

Along an airway, to find where you are in relation to an intersection, you will already know the bearing to station (BTS), because it will be on the map. Using the formula:

$$RB = BTS - MH$$

you can find what the needle will indicate when you get there.

Time to Station

As with the VOR, note the *seconds* taken to go through a number of degrees on the relative bearing, and divide the time just noted by the number of degrees gone through to get the time in minutes.

Then use groundspeed (or TAS in emergency) to find your distance.

Tracking

To find an intercept heading, just add or subtract the intercept angle to the track you wish to establish, as with an airway. It's common to use 90° inbound and 45° outbound, but use whatever ATC and circumstances (or exam questions) dictate (30° is nice). Note the track, and add or subtract your heading, as appropriate, to get the expected relative bearing when on track, which you will be when the needle of a fixed card points to it. With an RMI, just watch the needle.

When drifting, the needle will always point to the side of the aircraft the wind is coming from, so corrections inbound should always be made that way, ensuring that the needle actually goes to the *other* side of the longitudinal axis once a corrected heading is established. For example, if you want to track 090°, and the wind is coming from the right, to be on track you want to end up in a situation where the heading is an equal amount of degrees the other side of the lubber line as the needle is, such as a heading of 110° (plus 20 of the lubber line), looking for a 340° relative bearing (minus 20 of the lubber line).

Or, to use another example, for a track of 090°, your heading might be 070° while the ADF needle points to 110° (heading minus 20°, looking for plus 20° from the needle). If you are going the same way as your track, the needle will tell you which way to go. If it is on the left, your track will be on the left, and *vice versa*. Just turn whichever way until the needle reads the desired intercept on the opposite side. A good ploy is to allow the drift to happen until you get a positive reading, say 10° port, double it the other way (20° starboard), and when you are back on track, reduce by half (i.e. 10° in this case) to hold it.

When tracking outbound, however, you want to end up with the needle on the *same* side as the wind, so, although you are still looking for the plus 20, minus 20 equation, the

needle would be pointing at 160° RB. When you make your initial turn, the needle looks like it's going the wrong way, but it's something you get used to.

In short, if the pointy end of the needle moves to the right of a line between 0° and 180°, fly right, as drift is to the left, and vice versa.

DME

Distance Measuring Equipment is also UHF-based, between 962 and 1213 MHz. It is actually *secondary radar*, which measures the time difference between paired pulses being sent from the aircraft, and being received back on different frequencies, 63 MHz away (there are 126 DME channels). In other words, the aircraft is the first to transmit on UHF, then the DME transmitter returns the signal, plus 63 MHz. Two frequencies are used because, otherwise, the first pulse received would be the ground return from below the aircraft. Similarly, the ground station could self-trigger from other sources, such as those being bounced off a building. *Jittering* is used on the PRF so the DME's own pulse can be identified.

Instruments in the cockpit will not only show your distance to a station, but will calculate the rate of movement and display the groundspeed (just multiply the distance flown in 6 minutes by 10 if yours doesn't). It is normally based with a VOR or TACAN and has a range of about 200 nm, ± 6, with an accuracy better than ½ nm or 3% of the distance, whichever is the greater. The reason it's not completely accurate is because the distance measured is the *slant range* from the station, and not from your equivalent position on the ground, although at long distances and lower altitudes, this will be minimised.

GPS

The *Global Positioning System* was originally set up by the US military, using 24 satellites orbiting every 12 hours to give extreme accuracy at a very much reduced cost compared to, say, INS. Although they have guaranteed to keep the system running for the foreseeable future, in National Emergencies it may be unavailable, which is why it is still not acceptable as a sole means of navigation without traditional systems as backup (that is, it can be used as a primary *aid* in some areas).

It will give altitude information as well as location, but 3D readouts require 4 satellites. *Selective Availability*, where there was a deliberate fudging of the signal (by dithering the clock signals) to make it less accurate for non-military receivers, is now inoperative. However, for exam purposes, C/A (*Coarse Acquisition*) Code is made up from civilian signals on the L1 frequency, which is 1575.42 MHz. P (*Precise*) Code is for the military and broadcast on L1 and L2. Without C/A, accuracy is to 30m, otherwise it is 100m for about 95% of the time. *Differential GPS* uses a ground station within 70 nm to increase accuracy to within 1-3 metres.

Satellites are kept in line by an atomic clock at a ground master station, and aircraft use crystal oscillators. *Code matching* with a *time delay* removes *clock drift*, which is the system's major error.

At least 5 satellites should be visible at any point over the Earth at any time, though you could get a problem flying through the odd ravine way up North. The readings are referenced to a mathematical model (that is, an imaginary grid system), the WGS-84 ellipsoid; they can be converted to other models inside most receivers.

The *mask angle* is the lowest angle above the horizon from where a satellite can be used. Signals from it contain the time (from four atomic clocks) and its position, plus error correction. From this, the *pseudo-range* of the satellite can be computed, which is called that because it was not directly measured, but calculated, as a result of which it is subject to error, particularly delays as signals pass through the ionosphere.

For each satellite involved, the pseudo range is added to the *ephemeris* (or exact position in space), and triangulation used to figure out the receiver's position. As mentioned above, you need 4 satellites for a 3D position, which includes height above Mean Sea Level, in other words altitude. Such height readouts should not be used for navigation by themselves, as they are referenced to the WGS-84 spheroid, so the current altimeter setting should be put into the receiver if you want to use them (the techies at Garmin say the term *Mean Sea Level* is used generically, as the difference is close enough with the accuracy available).

RAIM stands for *Receiver Autonomous Integrity Monitoring*, for making sure satellites are working properly, which needs an extra satellite. For the bad signal to be isolated, you need one more than that (in fact, the 5th satellite is for checking errors, while the 6th is a stand-in).

Although it is tempting to use GPS all the time, remember that it is electrical, and therefore reserves the right to go offline at any moment, without warning. The antenna in a GPS is live as well, and equally liable to stop working.

Use it, by all means, but you should *always* have an idea of your approximate position, just in case it fails, or you go out of satellite range, so you don't get the embarrassment of having to explain to your customer just why you are landing to find out where the map is. It's just part of being a professional.

A GPS may have a database of airspace and frequencies inside – although not so important for VFR use when you carry a Pooley's, it is still the mark of a professional to keep it up to date.

Errors

- Satellite clock drift

- Ephemeris (position)

- Propagation delay

- Receiver noise

- Multi-path reflection

- C/A Selective Availability

Radar

The use of radar improves aircraft spacing and improves safety - the word stands for *Radio Direction and Ranging*. It works on the basis that radio pulses will reflect off objects and a proportion of the energy will return to the transmitter. You can calculate the distance between them because the speed of the radio wave is known, and the direction the aerial was pointing in at the time will supply the bearing. The "blips" representing the objects are displayed on a *Cathode Ray Tube* (TV screen) and an air traffic controller can see the relative positions of aircraft reflecting pulses:

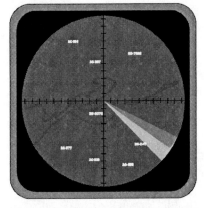

The word *pulses*, mentioned above, is the operative word – short bursts of electromagnetic energy are mixed with relatively long periods of silence (in electronic terms, this means somewhat less then a thousandth of a second).

This is known as *primary radar*, and it has a few limitations. First of all, radio signals weaken over distance and, since they have to make two journeys, the range of a target is necessarily limited. Secondly, the blip is quite large, and aircraft very close together cannot be distinguished. Finally, radio waves can be bent by the atmosphere or screened by objects, such as mountains or buildings, and different aircraft return signals differently, in terms of shape or surface.

Secondary Surveillance Radar

This is a development of a system introduced during the Second World War called *Identification Friend or Foe* (IFF), which was supposed to distinguish between friendly and enemy aircraft (friendly aircraft had a small transmitter that gave a distinctive periodic elongation to the blip on the screen).

SSR improves on the primary radar mentioned above by using secondary equipment to provide more information, hence the name. Participating aircraft carry a *transponder* (which stands for *transmitter/responder*) that receives the *interrogation pulse* from the transmitter, superimposes information on it and sends it right back on another paired frequency. This means, first of all, that the range of operation can be doubled immediately, and, secondly, that the blip on the screen can be made much smaller, together with information that makes it more easily identifiable to ATC, because the pulses can be coded. Computer trickery can provide predicted tracks and collision warnings, amongst other things.

A controller for example, will give you a number to *squawk*, which you dial up on the transponder in your aircraft, and which will appear next to your blip with a height readout, depending on the transponder.

There are standard numbers to squawk, when not otherwise instructed, and these are:

- 0000 – malfunction

- 0030 – lost

- 0033 – parachute dropping

- 2000 – from non-SSR area

- 7000 – conspicuity code

- 7004 - aerobatics

- 7007 – open skies

In emergency, squawk:

- 7500 - Hijack

- 7600 – Communications failure

- 7700 - Emergency

You will be given details of other traffic, on the clock system, such as "fast mover at 6 o'clock", based on the track seen on the radar screen.

When changing squawks as instructed, take care not to dial up the emergency ones by mistake, and do not switch the transponder to standby during the change to avoid it, as this will remove your display from ATC's screen.

A Mode A transponder is the regular variety, which just displays the code you select in the aircraft – you get this by turning the switch to ON.

A *Mode C* is directly attached to an encoding altimeter (or, more precisely, an *altitude digitiser*, which selects a different code to that selected in the window), but only Pressure Altitude information is sent from the aircraft – the conversion to local pressure, if required, is done inside the ATC computer. ATC will not be able to see changes when you move the subscale. Mode C is selected by switching to ALT.

If a transponder fails during flight in a mandatory area, you may go to the next planned destination, then complete an itinerary or go to a repair base, as permitted by ATC.

It is possible to enter controlled airspace without the required equipment, but ATC must be asked first. It is always subject to traffic.

When asked to *squawk ident*, your return becomes brighter for a short time, for positive identification.

Pilots report to the nearest 100', and ATC confirm to the nearest 200'.

VDF

ATC can get a bearing for you to steer (QDM) to get to their location from your transmissions, when using *VHF Direction Finding* equipment. Being based on VHF, it is subject to the usual limitations (see above), so the higher you are, the better the results you will get. You need to transmit for a few seconds for a bright line to spread from the centre of a screen to the outside which is marked with compass bearings.

- QDM - magnetic bearing *to*

- QDR – magnetic bearing *from*

- QUJ – true bearing *to*

- QTE – true bearing *from*

A VDF letdown exists where ATC give you QDMs, and you work out the headings to steer (a QGH is a military letdown, where ATC give you the headings to steer).

Older equipment will use a cathode ray tube on which the line appears (rather like a radar sweep) pointing to where your transmission is coming from. More modern digital equipment uses a circle of LEDs at 10-degree intervals, which will show the same information, with a digital readout in the centre. The controller can store the last transmission, in

case of being busy with something else at the time. Radar can help by giving distance information as well.

Accuracy comes in three classes, A, B or C. Alpha is $\pm\, 2°$, B is $\pm°5$ and C is $\pm 10°$. You will only be told the classification if the bearing is B or C.

Radio Procedures

Like aircraft, pilots need radio licences in order to use the airwaves. This is normally a separate licence, which needs you to pass an exam before it is granted. Although your use of the airwaves is limited (it's a subset of the amateur radio regulations), you still need to know the phraseology so that other people don't suffer.

Due to licence restrictions, aircraft equipment is meant to have as few controls as possible, including displays, so some frequencies may not be completely shown (122.075, for example, comes up as 122.07).

The *Squelch* quietens down the output when no signal is being received, so you don't get continuous earfuls of white noise. A signal coming in cancels this and activates the audio (a variable squelch merely determines the signal level when this occurs). The correct procedure with the Squelch control, therefore is to rotate the knob until the hiss just stops, and leave it there, although it is true to say that this will hide a weak signal, so lifting the squelch may help in this case.

Although the phraseology can be a bit longwinded (*day-se-mal* for decimal, for example), and you may feel a bit stupid pronouncing some of the words, remember they are that way to reduce ambiguity.

Naturally, a continuous listening watch should be maintained at all times, as a matter of airmanship, but especially when transiting controlled airspace during notified hours of watch. You must report your position and height on entering and just before leaving an ATZ.

There are one or two other points about radios that aren't often taught properly during training. The first is to wait a split second to speak after pressing the transmit button, which gives all the relays in the system a chance to switch over so your message can get through in full, that is, not clipping the first bit. Secondly, whenever you get a frequency change en route, not only should you write it down on your Nav Log, but change to the new frequency *on the other box*, so you alternate between radios. This way, you have something to go back to if you can't get through on the new one for whatever reason (although it is appreciated that this could create difficulties with two station boxes which must be switched every time).

You should use the full callsign on initial contact with ATC, but you can subsequently use any abbreviations they make.

If you need to make a correction, say the word "Correction" followed by the last correct word or phrase before continuing.

Numbers should generally be spoken individually, except for the words hundred and thousand where they occur as round figures. 100, for example, would be "One Hundred", but 165 would be "One Six Five".

Usually, with regard to time, you transmit the numbers relating to minutes (e.g. "arriving at 45"), but this only relates to the *current hour*. If there is any possibility of confusion, or you mean another hour, include the figures.

Standby means "wait to be called". *Affirmative* means an agreement, but the word was changed (at least in Europe) some time ago, to *Affirm*, to reduce the possibilities of it being confused with *negative* if only the last part of the word was heard. In the same vein, you only use the word *Takeoff* when cleared, or cancelling a takeoff clearance. For an abandoned takeoff, use the word *Stopping*.

Roger means that the last message has been received (even if you didn't understand it!).

The readability scale is:

1 – unreadable

2 – readable now and then

3 – readable with difficulty

4 – readable

5 – perfectly readable

The order of priority of radio messages is:

- Distress
- Urgency
- Direction Finding
- Flight Safety
- Meteorological
- Flight Regularity

ATC

At smaller aerodromes, some of these may be combined.

Ground Control

The Ground Controller handles all movements on the manoeuvring area, including aircraft and vehicles, and possibly start clearances (departure clearances given by Ground are *not* clearances to takeoff!). Typically, you would be talking to Ground up to the holding point, and afterwards when landing.

Tower

For traffic close to the aerodrome, including the circuit. After takeoff, you may be asked to change to *Approach* (below), but, more typically, you will stay with the Tower until clear of the area.

Approach

Sometimes known as *Radar*, these controllers sit in a darkened room in front of radar screens, so have no visual contact with the traffic they are dealing with (don't worry, they are fed frequently).

Radio Failure

Essentially, comply with the last clearance, which hopefully included permission to land or clear the area. If you don't need to enter controlled airspace, carry on with the plan, maintaining VFR as necessary; don't enter it even if you've been previously cleared. If you must do so, divert and telephone for permission first. If you're already in controlled airspace, where clearance has been obtained to the boundary on leaving, or the field on entering, proceed as planned. If in doubt, clear the zone the most direct way as quickly as possible, avoiding airfields.

The military have a system of flying a left or right-handed triangle pattern that can be seen on radar, although it's usually only used if you're lost as well as having a duff radio. Use it as a last resort, though, because ATC have other things to look out for than possible triangles. If they do recognise your problem, they will send up a shepherd aircraft to formate on you and bring you down, so remain VMC if you can, and as high as possible so radar can see you better. If you can squawk Mode C, do so, because that will give a height readout to work with. If you can only receive, fly in a right-handed pattern for a minute (over 300 knots, make it two). Fly at best endurance speed and make each 120 degree turn as tight as possible. If you can't transmit either, do the same, but to the left.

RT Emergency Procedures

Always declare an Emergency, even if you have to downgrade it later.

Distress

The Distress call (or "MAYDAY") is used when threatened by *grave and imminent* danger and in most urgent need of *immediate* assistance (like when your a single engine fails), you can use the letters SOS in Morse Code (... --- ...), or the spoken words MAYDAY, repeated three times, followed by any relevant details, like your position. You can also fire rockets or red lights at short intervals, with parachute flares.

The official frequency is 121.5 MHz.

If and when the threat is over, the Distress call must be cancelled by notification on ALL frequencies on which the original message was sent. To cancel a MAYDAY:

- State the word MAYDAY once

- Say ALL STATIONS three times

- Aircraft ID

- Station called

- Time

- Name of station in distress

- DISTRESS TRAFFIC ENDED

- Station called

- OUT

If you hear a distress call, you should record the position of the aircraft in distress, and the bearing, inform ATC or an RCC, then proceed to the position.

Once there, if a rescue is in progress, do not interfere without checking with the controlling aircraft.

Urgency
The Urgency call (or "PAN") spoken three times, indicates a very urgent message concerning the safety of a ship, aircraft or other vehicle, or of some person on board or in sight. If you just wish to mention you are compelled to land, but don't need help right away, switch the landing lights and/or navigation lights on and off in an irregular pattern.

The official frequency is 121.5 MHz.

Some Questions

1. Why does attenuation occur?

2. As frequency increases, does the dead space become larger or smaller? What happens to the skip distance?

3. What is the Maximum Useable Frequency?

4. What is QDM? QDR? QTE?

5. When will the VOR flag appear?

6. If you are tracking towards a VOR on the 180° radial, and drift is 15° to port, what is your heading?

Some Answers

1. The circumference of the wave front increases, the Earth's surface absorbs some of the energy, and so does the ionosphere.

2. It gets larger. The skip distance increases.

3. The frequency that causes the first returning sky wave to fall just short of the receiving station.

4. QDM is the magnetic great circle bearing of a station from an aircraft, or the great circle heading to fly to the station in still air conditions. QDR is the reciprocal of the QDM, often referred to as a *radial*. QTE is the true great circle bearing of an aircraft from a station.

5. When the transmissions are faulty, the aircraft is out of range or otherwise out of the signal path, or if there is a power failure.

6. 015°. You are on the 180° *radial*, so you are pointing *towards* the VOR. Just add the drift. (in an exam, you would probably be asked to draw a diagram of your position with reference to more than one beacon).

Navigation

Navigation involves taking an aircraft from place to place without reference to the ground, except, perhaps, for checking you've got the right destination! To do this, a system called *Dead Reckoning* is used, which is actually short for *Deduced Reckoning*, based on solving a triangle of velocities, discussed below.

First, however, we need to get acquainted with the Earth, which is not actually round, but flatter at the Poles than at the Equator. For our purposes, though, and the mathematical models inside the average GPS, it is a sphere.

To help find your position, a series of lines is drawn from Pole to Pole through the Equator, called *lines of longitude*. They may also be called *meridians*, when split in half, and by convention are drawn for every degree you go round the Equator. Also, by convention, they start at Greenwich, in London, England (with the *Prime Meridian* at 0°), and are calculated to 180° East or West:

The opposite side to any meridian is its *anti-meridian*.

Since the Earth takes 24 hours to spin on its axis, 15 lines of longitude represent 1 hour, and it is noon when the Sun is overhead any particular meridian. The spinning is anticlockwise when viewed from the top of the Earth, so the Sun will appear to rise from the East and set in the West.

however, is not enough, since you could be anywhere on it, so more imaginary horizontal lines are drawn, parallel to each other, North and

South of the Equator, up to 90° each way, called *lines of latitude*. Now you can get lost more accurately! The latitude of any point is the arc of the meridian between the Equator and the *parallel* through the point:

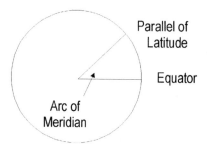

Parallel of Latitude

Equator

Arc of Meridian

Latitude lines are always parallel to each other, whereas longitude lines (meridians) converge. They are also fixable by natural means - the Sun, Moon and planets pass over the Equator, for example, and the tropics of Cancer and Capricorn represent the limits of the Sun's travel North and South as it rises and sets every day (it sets further South each day, until, on December 21st, it stops for three days to go North again. On June 21st, it stops going North to go South).

Ptolemy had plotted some sort of lat & long system by 150 AD, but he used the Canaries for the Prime Meridian, which has also been at the Azores, Cape Verdi, Rome and Paris, to mention but a few (it wouldn't surprise me to learn the Ancient Egyptians used the pyramids). Eventually, it was placed in Greenwich, because King George was a keen astronomer.

When giving position, latitude is always given first, as in 45°N, 163°W. The distance between *parallels of latitude* is 60 nautical miles, because 1 nautical mile is the distance covered by 1 minute of latitude, but it varies slightly between the Poles and the Equator because the Earth bulges in the middle (6080 feet is used, but it is actually only correct at 48° latitude). One minute of longitude, however, will only be 1 nm at the Equator, due to *convergency* (the distance between meridians gets smaller toward the Poles).

Since we take 24 hours to go round the Sun, in one hour, we move through 15°, or we take 4 minutes to go through 1°. Similarly, in 1 minute we transit 15 minutes, or take 4 seconds to go through 1 minute.

Just to remind you, a *degree* is split up into *minutes*, which in turn are split into *seconds*. Also, meridians diverge in the Southern Hemisphere.

Great Circles & Rhumb Lines

Great Circles have planes that go through the centre of the Earth, or, in other words, are circles whose radius is that of the Earth, so the definition includes lines of longitude and the Equator. Since meridians are half lines of longitude, they are *semi-great circles*. Although they are the shortest distance between two points, the angle created when they cross meridians changes (convergency again), so your course is continually under review.

Rhumb Lines, on the other hand, cut each meridian at the same angle, but they are not straight (they are concave toward the nearer Pole), so are longer in distance. All lines of latitude are rhumb lines, but the Equator is a great circle as well (lines of latitude are also *small circles*).

Direction

This is always expressed with reference to *True North*, that is, ignoring any magnetic effects from the earth itself. A *bearing* is the clockwise angle between North and any line between two points.

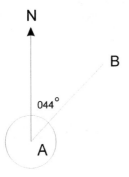

In the above case, B is on a bearing of 044° from A, in relation to North. The opposite is the *reciprocal*, quickly found by adding or subtracting 180°, that is, 224°.

Magnetic Bearings

One problem is that a compass does not point towards True North, but *Magnetic North*, since the Earth generates its own magnetism – and the two Norths (or Souths) do not coincide at their respective Poles. The next is that the magnetic force is not constant over the globe – it may be varied by local deposits of metals under the ground, for example, and bend the magnetic flux lines. The way to Magnetic North will therefore vary from place to place. In addition, the lines of force will be vertical near the poles:

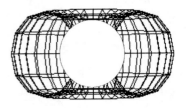

The North Magnetic Pole was discovered by Soviet explorers to be the rim of a magnetic circle 1000 miles in circumference:

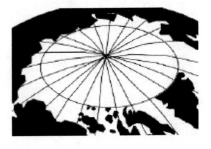

On a map, which is drawn initially for True North, there is a dotted line called an *isogonal* that represents the local magnetic variation to be applied to any direction you wish to plan a flight on:

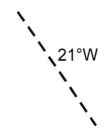

It is westerly where the variation is to the left of the meridian, and Easterly when to the right. It also changes every year, since the magnetic pole moves East, about one degree every six years.

Isogonals are accurate worldwide to ±2°. Magnetic variation, therefore, is the angle between True North and Magnetic North. An *agonic line* exists where magnetic variation is zero, or where they are both the same. There's one near Frankfurt, running North/South. The line of zero dip (the Equator) is the *aclinic* line.

The phrase to remember is *Variation East, Magnetic Least, Variation West, Magnetic Best*, that is if the variation on your map is, say, 21° West, the final result should be 21° *more* than the true track found when you draw your line. If you travel over many variations, use an average about every 200 miles.

By convention, the North Pole is blue, and the South Pole is red.

Remember also that variation on a VOR bearing is applied *at the station*, and on an ADF *at the aircraft*.

Time & Time Zones

You know already that the Earth, together with 8 other planets, revolves round the Sun. 1 year is the time it takes to go once round, in the Earth's case 365¼ days (the odd quarters are consolidated every four years into one day in a *leap year*).

The Earth does not spin vertically, like a top, but is inclined. When the inclination points towards the Sun, the Northern Hemisphere days are long and the nights are short. The day when this is at its maximum value is the *Summer Solstice* on June 21 (*Solstice* is Latin for *Sun Stand Still*). The *Winter Solstice*, when the inclination is at its maximum away from the Sun is December 21. Days and nights are of equal length on the

Spring and Autumn Equinoxes, March 21 and September 23 (*Equinox* means *Equal Night*), because the spin axis is vertical to the Earth's orbit.

The Prime Meridian is the standard to which all local mean times are referred. Local Mean Time there used to be called *Greenwich Mean Time* (GMT), but is now referred to as *Universal Coordinated Time*, or UTC. The Greenwich day starts when the mean Sun transits the anti-meridian, and transits the Easterly ones before it reaches Greenwich. The local mean time in those places will therefore be ahead of UTC, and that of those West will be behind. When doing calculations, revert everything to UTC first, and don't forget the date! The *International Date Line* is where a change of date is officially made, being mainly the 180° meridian which bends to accommodate certain islands in the South Sea and parts of Siberia.

Deviation

We saw above that the magnetism from the Earth will vary the direction displayed by a compass. The aircraft's magnetism, created from large amounts of metal mixed with electric currents, will do the same thing, called *deviation*, which is applied to the magnetic heading to get *Compass North*.

The phrase here is *Deviation West, Compass Best, Deviation East, Compass Least*. This means that if the deviation is to the left of the magnetic North, the difference should be *added* to the course to get the correct magnetic heading.

Deviations will be displayed on a small *correction card* next to the compass, and are obtained after a *compass swing*, a complex procedure normally done by an engineer. There will be an area on every aerodrome well away from buildings, etc. set aside for this purpose. Allowing for deviation is called *compensation*.

Maps & Charts

The words map and chart are nowadays used interchangeably but, officially, a chart will show parallels and meridians with minimum topographical features, and be used for plotting. A map will show greater detail of the Earth's surface.

The point about them both is that their representation of the Earth's surface is only accurate within a relatively small area, since you are trying to show a 3 dimensional object on a 2 dimensional surface. The further from the *centre of projection* you go, the more the distortion is but, to all intents and purposes, it can mostly be ignored. You can see the problem if you flatten a globe:

There are many ways of compromising for this, and each suits a different purpose, so lines drawn on maps based on different projections will not cross through the same places.

The quality of *orthomorphism*, that all charts should strive for, means the scale must be correct on all

directions within a very small area. In addition, parallels must always cross meridians at right angles. Otherwise, no chart is perfect, as you will find when you fold them:

Lambert's Conformal
Imagine the Earth with a light shining at the centre, then place a cone on the top. Where the cone meets the earth, the shadows of the land formations will be accurate, but will be out of shape the further North and South you go.

This is the *conic projection*, the basis of the *Lambert Conformal*, and is what most of the charts used today are based on, as the meridians will be straight, even if they converge towards the North:

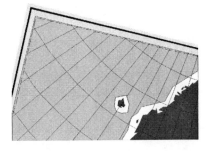

Great circles are assumed to be straight lines (actually they are *very* shallow curves), and rhumb lines will be curves concave to the nearer pole. Johannes Lambert overcame the problem of scale expansion in the 18th century by pushing the imaginary cone further into the Earth's surface, so it cuts in two places:

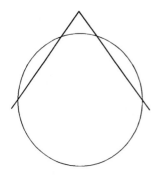

This gives it two *Standard Parallels*, or points where scale is correctly shown. To be sure, there is a slight contraction between them, but this is considered insignificant (1% or less) if two-thirds of the chart are between the Parallels.

Mercator

The *Mercator projection* does things differently. Instead of a cone, the Earth is surrounded with a vertical cylinder, touching at the Equator. Meridians now do not converge, so rhumb lines will be accurate, but distance between latitude lines increases away from the centre (not significant below about 300 nm, but always use the scale near the distance to be measured):

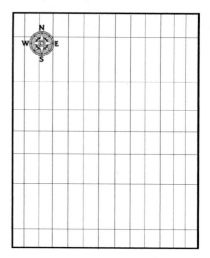

Again, shapes will be accurate where the cylinder touches the surface, but the distortion will be much greater the further away (as a point of interest, Mercator was the first chart to be used for maritime navigation in the 16th century). Since rhumb lines on this projection are straight lines, it follows that great circles must be curved, in this case, *concave to the Equator*, that is, the rhumb line is always nearer the Equator.

The rhumb line looks shorter than the great circle because of scale expansion. The relevance of this lies with plotting radio bearings, because radio waves take the shortest way (e.g. great circles), so long distances need the conversion angle to be applied to plot them as straight rhumb lines – in fact, an ABAC scale on the chart will do this for you. Complications also arise from whether the plot is done at the aircraft (ADF) or the station (VOR/VDF), but we won't go into that here.

The Mercator projection is the one mostly used for plotting charts, as constant headings are easier to use.

Transverse Mercator

This is a horizontal cylinder projection, and a straight line still represents a great circle. The *Central Meridian* (CM), where the cylinder touches the sphere, coincides with the relevant latitude, so True North and Grid North are the same along it. However, because rectangular grid lines are drawn based on the CM, moving East or West means applying some sort of grivation (see below). A scale factor also has to be applied as you move around the map to convert ground distances to measured distances. To reduce this, the projection uses two North-South lines with a scale factor of 1, so in the centre the correction is less than 1 (0.9996 for the UTM), while the outer parts have it greater than 1.

The WAC at 1:1,000,000 and the VNC at 1:500,000 are based on a two-parallel Lambert Conformal Conic, whereas the larger scale VTA at 1:250,000 uses Transverse Mercator (the Vancouver VTA covers a bit over 2° of longitude).

The Transverse Mercator's advantages include accuracy (over small areas, at least). Wide countries are split into zones usually no wider than 6° of longitude, at which point the distortion becomes unacceptable.

Universal Transverse Mercator

UTM is a metric grid system based on Transverse Mercator, designed by the US military, using 60 6° longitude zones and 20 8° latitude bands between 80° S to 80° N, giving 1200 areas overall. Longitude zones are numbered 1-60 starting at 180°W. Latitude bands are lettered from C (not I and O) Northwards from 80°S. Each 6° by 8° area has its own grid, based on 100,000 m squares. Each column and row is lettered, and when the numbers are used, Eastings are given first.

Polar Stereographic

These charts are used in polar regions, because the others cannot cope with convergence that well. To get the details correct, the paper is held flat over the top of the Pole and the imaginary light projected straight up from the centre of the Earth to it:

On these, rhumb lines are not the shortest course, and you must use great circles instead. Since they are straight lines, meridians are crossed at different angles.

The Arctic

Up there, it's darker for longer and there are fewer navaids. The compass begins to get unreliable, and there is increased deviation due to the aircraft's own magnetic field.

Scale

Assuming a constant scale, the ratio between distances on a map and the Earth's surface is expressed as a *scale* based on the map's size. For a scale of 1:500,000 (commonly referred to as a *half-mil*), one inch on the map is equal to 500,000 inches on the Earth. There are 63,360 inches to the mile, so an inch on a half-mil map is 7.89 statute miles.

You can tell which chart has a larger scale by looking at the *representative fractions*, obtained by dividing chart distance by Earth distance. Thus, a chart distance of one inch divided by its Earth equivalent of 13.7 nm would be a 1/1000000 map, and of a smaller scale than a 1/500000.

A "one-inch map" means one that uses one inch for one mile. A "quarter inch map" has 4 miles to the inch (about the length of the distance between the joint on your thumb and the tip, for quick reference). Not everything on the map is done to this scale; if it were, you would hardly see roads and railways, so they are artificially expanded to be visible. The centre of any object is its actual position.

The faster your aeroplane flies, the less time you have to check the map, so those made for high and fast flight, or for instrument flying where you can't see the ground anyway, will not have many ground features marked on them.

If you need to find out what Earth distance is represented by a chart distance, multiply the chart distance by the scale. For example, if asked what distance is represented by 25 cm on a 1/1,000,000 chart, multiply 25 x 1,000,000 to get 25 million

centimetres. Divide that by 100 to get metres (250,000), then divide by 100 to get 250 km.

Relief

Information about high ground is given in various ways.

Contours are lines on a map joining points of equal height (or elevation) above sea level, so they are similar to isobars, in that the closer they are together, the steeper the slope they represent:

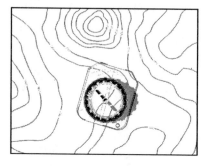

Some maps may give different colours or shading to various layers to make things more obvious, known as *Layer Tinting*.

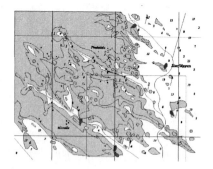

Spot Heights show the elevation of prominent peaks with small dots. The actual height will be shown next to it. The highest one will be distinguished in some way.

Otherwise, expect water to be blue, woods to be green, and railways and power lines to be black.

Speed & Distance

A *knot* is a measure of speed, that is 1 nautical mile per hour (it was originally measured by allowing a long rope to stretch out behind a ship – knots were tied in it at regular intervals, hence the name). 1 *nautical mile* (nm) is taken as 6080 feet, slightly more than a *statute mile*, which is 5280 feet.

A kilometre is 1000 metres, and is a 1/10,000 of the distance between the Equator and either Pole. 8 km equals 5 statute miles.

Triangle of Velocities

In flying between point A and point B, you will only get there by just pointing the nose in the right direction if there is no wind, or if it is exactly on the nose or tail. This is very rarely the case, so your aircraft would drift off course, according to the wind's direction, if you did nothing to correct it. In other words, you would end up a certain distance left or right of the original target (in the early days of the North Sea, when navaids weren't around, pilots would build in a slight error to their calculations, so that they would know which side of the rig they were just in case it all went wrong).

The smart thing to do would be to make a heading correction towards the wind's direction to maintain a straight track. This, unfortunately does two things. Firstly, the body of the aircraft is inclined more sideways to the track and, secondly, groundspeed is reduced, because

some of the energy from the engine is used in keeping it there.

You can see from the above that the speed of the aircraft through the air is not necessarily the same as its speed over ground - if you are flying into wind, you will go slower relative to the surface, and faster if the wind is behind you. An aircraft in flight is affected by the wind both along its axis and from the side, or from a head/tail or beam component.

You work out what the wind's effect on your trip will be by getting the forecast winds from the flight planning office, and working out a combination of three sides of a triangle, called the *triangle of velocities*, because a velocity expresses a combination of speed and direction, and we are concerned with those of your aircraft, the wind and the difference between them.

First of all, a few definitions:

- *Track*. The path the aircraft intends to follow over the ground, represented by the line on a map from one point to another (*Track Made Good* is the actual path – the difference between them is *Track Error*).

- *Heading*. The direction in which the aircraft is pointed, according to its compass, with reference to True North, because Track (above) and W/V (below) are.

- *Wind Velocity*. The speed and direction of the wind, based on True North. The faster your aircraft, the less its effect.

- *True Air Speed* (TAS). The speed of the aircraft relative to the atmosphere, not necessarily the same as that indicated, and not necessarily the same as.....

*   *Ground Speed*, or the speed of the aircraft over the ground.

*   *Drift*. The difference between track and heading due to wind.

*   *Air Position*. The position the aircraft would have reached without allowing for the wind.

*   *DR Position*. The calculated position of the aircraft.

*   *ETA*. Estimated Time of Arrival.

*   *Fix*. Definite confirmation of position by ground observation, radio aids or astro nav.

The velocity of an aircraft in flight will therefore consist of its heading and airspeed, the former usually expressed with reference to True North:

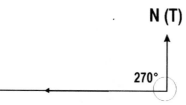

In the diagram above, the heading is 270°(T) - the single arrow is the symbol for the heading vector, pointing the right way, of course. When plotting, a scale is used, so if the heading vector were 3 inches long, at 50 kts to an inch it would equal an airspeed of 150 kts, or the *air position* after one hour of flight. If we added the wind speed and direction, the resultant between them would represent track and groundspeed, also to scale:

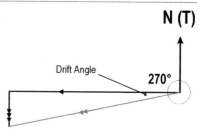

In this case, the wind vector is half an inch long, meaning 25 kts, coming *from* the North. Joining the ends would therefore show your ground position after one hour, and your track and groundspeed, after measurement (you will have deduced already that two arrows are used for the track and three for the wind – the track arrows always go in the opposite direction to the other two). The *drift angle* is the difference between track and heading, *from* the heading *to* the track, in this case about 10°, so the track is 260°.

The above diagram shows what would happen if you simply pointed the aircraft nose towards the West – you would drift to Port for the amount indicated. If you wanted to arrive over the intended destination, you would have to point the nose to the right (i.e. Starboard) enough to counteract the drift.

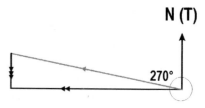

All you need to do is draw the same wind vector on the opposite side of the line, and measure the length and angle of the new line to find out what heading to steer (280°). Don't forget the variation and deviation (See *Instruments*).

Dead Reckoning

This involves the calculation of your best known position without navaids or visual fixes. In essence, it involves drawing the equivalent triangles of velocity you would create on your Dalton computer (see below) on a map, although it is important to grasp that the triangle's purpose is more to do with finding directions and speeds rather than finding a position. As mentioned above, the lines you draw will be to scale, so one 3" long at 50 miles to the inch would represent 150 kts. When climbing and descending, take the mean TAS for the leg, and mean wind velocity.

If you happen to fly over an object that can easily be identified from a map, you have a *fix*, which can be used to find what the real wind is, and your actual groundspeed. Simply connect a line from your air position to the fix, and measure the resulting line between them (the *wind vector*). The line between your start point and the fix would be the *Track Made Good*, which could be used to solve the above problem on the computer.

Remember that these velocities go together: *Heading & Airspeed, Track & Groundspeed, Wind Direction & Speed*. However, you will be involved in finding mixed pairs, such as heading and groundspeed, rather than the combinations mentioned above, because you start with a mix in the first place (you usually know the airspeed and track already).

Given any four of them, you can figure out the others purely by measurement, but you can do this mechanically with the flight computer, or whizzwheel, described in the *Flight Planning* chapter.

The 1 in 60 Rule

This is a common method used in solving tracking problems, based on *tangents*, which, if you remember from Pythagoras, can be found by dividing the length of the opposite side of the angle to be found by the adjacent side. Or, in terms of aviation, dividing the distance off track by that of the desired track. We needn't go into the proof here, but you can end up with a formula:

$$\text{Error} = \frac{\text{Distance Off} \times 60}{\text{Distance Gone}}$$

So if, after flying for 40 nm, you are 8 nm off track, your track error angle would be:

$$\text{Error} = \frac{8 \times 60}{40}$$

or 12°. This would be doubled the opposite way to get you back on track, then applied as a single figure to keep you there (applying the correction once would make you parallel the original track).

To track directly to the original destination, you would need an extra bit, called a *closing angle*, which you can find by altering the formula:

$$\text{CA} = \frac{\text{Distance Off} \times 60}{\text{Distance To Go}}$$

Add the combination of closing angle and track error to the heading the appropriate way.

If you were intending to track along a VOR radial, and found you were actually on a different one, you can use the 1 in 60 rule to see whether you were still inside the airway. If the centreline was 045°, and you were on the 040° radial, you would be off track by 5°. If the DME says you are 45 nm away, it is a simple calculation:

```
Dist Off = TE x Dist Gone
                  60
```

The answer is 3.75 nm, so you are OK.

Summary

The 1 in 60 rule means that every 1 degree off track represents 1 nm for every 60 travelled.

If you just want to parallel the track, alter course by the track error in the appropriate direction. To go to the destination, add the closing angle.

To get back on track, alter course by double the track error. Once there, the original track plus or minus the track error will keep you there.

Be aware, though, that the time to regain track may be more than that used to create the error in the first place, and that these rules are approximate, because altering heading changes the relationship of the wind to your machine.

Departure

For the most accuracy when learning to fly, it is best to get to a safe height over the departure point, *then* set course, making sure you pass the start point in the cruise. Although this means you won't have to make separate calculations for the climb, with more experience, you will be able to climb away on course directly from the circuit, making the proper allowances.

En Route

Accurately fly your planned heading for six minutes, note your position, then compare it to the map, and the 10 nm intervals you drew on it. Whatever distance you have flown in six minutes multiplied by ten is your groundspeed, so 10 miles in 6 minutes is 100 kts. You now have either confirmation of planned groundspeed or a new one, so you can create a new ETA.

Check your DI against the compass every 15 minutes or so. Also, check your fuel state against your progress, noting large reductions in particular, as they may indicate that you have left the fuel cap undone, or that you have a leak.

The Circle of Uncertainty

Assuming you have flown as accurately as possible, and the wind velocity was accurately forecast, and you made no mistakes in your flight planning, you should find yourself pretty much on track throughout the flight. However, life is not always like that, and once in a while you may find yourself unsure of your position, which is the technical term for being lost.

The circle of uncertainty is a way of trying to remedy this by allowing a percentage of error and drawing a circle of appropriate size centred on your destination. In theory, you should be somewhere inside it. Its diameter will very rarely be more than 10% of the track distance.

Map Reading

Get used to recognising ground features from the map and angles and distances between them. When identifying a fix, you need at least three ways of confirmation.

Also get used to not necessarily needing to know exactly which field you are over at all times – a common fault is too much accuracy when you start flying. Knowing you are so many miles in a particular direction from somewhere is good enough.

If operating inside a particular area, choose a prominent landmark and rotate round it, that is, keep an eye on your position in relation to it.

Check Features are prominent landmarks selected in the planning stage to look for during the flight.

The span of your hand is about nine inches, very useful for measuring distance. From the middle of your thumb to the tip is about an inch.

Some Questions

1. How far will you fly in 2 hours 38 minutes at a groundspeed of 364 knots?

2. If an aircraft flew 60 nm in 14 minutes with a TAS of 250 kts, what is the wind component?

3. With a wind velocity of 260/15 and a track of 296°T, what heading and airspeed should be flown to maintain a groundspeed of 120 kts on track?

Some Answers

1. 960 miles. Line up 60 against 364 on the outer scale, then look for the answer against 158 minutes.

2. Line up 14 on the inner scale against 60 on the outer, to read the groundspeed of 257 kts against the 60 triangle. Since this is more than the TAS, there is a 7 knot tailwind.

3. 292° at 132 kts.

Flight Planning

This may appear tedious in the early stages, but planning is actually around ¾ of a trip. The more you do, the more answers you will have to hand when things go wrong and the better the trip will be, as any plan you have spent time over is better than one cooked up on the spur of the moment. If you get yourself into a little routine, the process will become speedier as time goes by.

Proper Planning Prevents Poor Performance.

Well begun, half done.

Points to remember are the weather details for the destination and alternates, plus takeoff alternates if you have to land back in a hurry (this could mean up to eight airfields), check the runways available (and crosswinds!) and NOTAMS, in case any aids are out.

Sometimes ATC have *preferred routings*. Check the altitudes on the chosen route, and ensure you have the performance to maintain them – adjust your all-up weight.

Always ensure you are using the same units, by which I mean don't mix magnetic and true headings and wind directions. Either apply magnetic to everything before you start, or work it all out in true and apply the variation at the end.

Keep ATC informed of delays, otherwise SAR might be called out by mistake!

Fuel

Very few aircraft will actually take a full load of passengers and fuel, so you need to know how long it will take between two points, find out how much fuel it will take, *then* fit the passengers in. *Do not put the passengers in first and fit the fuel in afterwards!* Not unless you plan to stop en route, at least. Of all the things there is absolutely no excuse for in Aviation, running out of fuel is one of them! If you have to take less fuel, then you will have to stop and pick up some more on the way, or leave someone behind. If you take the same fuel anyway, you will be overweight, with not enough power in the engines to get you out of trouble, and *invalid insurance.*

Although not written down in the ANO, commercial pilots use these figures for fuel planning.

Helicopters

You should be able to fly for 20 minutes at normal cruising speed after reaching your destination. At night, it's 30 minutes.

Fixed Wing

The plan should be to arrive over the destination in a position to make an approach, overshoot and fly to an alternate, and still have enough to hold for 45 minutes at the alternate. Even then, you must still be able to carry out an approach and landing, and you need 5% for contingencies. Don't forget any unuseable fuel your machine may require.

Maps

When planning a flight, the first thing to do is to draw a line on the map to represent the track you wish to fly. Find the mid-point and mark it with a cross. Then mark the line at 10 nm intervals (or split the two halves into quarters), and draw dotted lines branching out at 10° from the origin. These may be left out once you know what you're

doing, but when learning they are very useful when calculating drift once you find the wind is different from that forecast and you need to recalculate on the run. The less you have to do in the air, the better, as your first priority is to fly the aircraft, and you don't want to start getting rulers out and spreading your map around in front of passengers.

The PLOG

The letters are short for *Progress Log*, or a sheet of paper which tabulates the details of a particular flight, used for flight planning and checking progress on the actual trip. In commercial companies, it must be kept as a record of the flight.

Once you've drawn your line on the map representing the desired track, you put its details in the appropriate boxes on the plog, work out the wind, obtain your intended heading and groundspeed, apply the magnetic variation, calculate the fuel required, fill 'er up and you're ready to go.

Below is a sample form, partly filled in with details of a proposed trip from Glasgow (GOW) to Inverness (INS). Notice that the Flight Level (or altitude, in this case) is *higher* than the Safety Altitude, which is the

Time	From	To	FL/ Alt	Safety Alt	TAS	W/V	Track T	Drift	Hdg T	Var	Hdg M	G/S	Dist	Time	ETA
1200	GOW	INS	4500	4300	90	180/15	007			-5·5			104		
Alternate											Totals ➝				

Time	From	To	FLt Alt	Safety Alt	TAS	W/V	Track T	Drift	Hdg T	Var	Hdg M	G/S	Dist	Time	ETA
1200	GOW	INS	4500	4300	90	180/15	007	+1	008	-5.5	013	105	104	60	1300
Alternate											Totals →				

higher of the highest *ground* within 5 nm of track, plus 1299 feet, or the highest *structure*, plus 1000 feet (in this case, I've taken the biggest blue figure in the lat/long boxes en route, off the half-mil map).

Otherwise, there's not much else you can usefully put in at this stage, so get your whizzwheel and see if you can fill in the rest, given that the wind velocity is 180/15. If you want to cheat, the picture above will show you what it should look like.

The figures for an alternate aerodrome have been left out for clarity, but you should always choose one and work out the figures for it in the same way, *before* you go - by definition, an alternate is for when you *really* need one, and there's never enough time to do things on the run.

When planning a trip with a lot of legs, my own preference, if there's room, is to leave a line between each one, in case you do have to change things, or you note any differences, such as wind velocity, and work out a new groundspeed.

Different flying clubs will have their own version of the above form, but this is most like what you will see in the exam. The only thing that's missing is a box for your fuel totals, but it's a simple slide rule calculation,

based on your flying time against fuel consumption.

The Dalton Computer

This is a device with a sliding scale through it, marked with drift angles and TAS arcs, with a frosted circular screen on which you can draw the business end of the triangle of velocities:

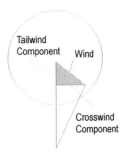

There is a dot in the centre of the screen, around which is a compass rose that can be rotated to bring your heading or track under the lubber line. All you need to do is draw in the wind vector to see how they all relate to each other.

Note: there will be an instruction book supplied with your computer, so the instructions given here will necessarily be brief.

The first thing to do is move the sliding scale to make your TAS appear underneath the dot in the centre of the frosted screen. Then rotate the screen so the wind direction lines up under the lubber line at the top.

Draw in a line vertically downwards from the centre dot equal to its speed in knots. Rotate the screen again until the track is under the lubber line. The end of the wind line will point to a drift arc and a TAS arc, which you just apply to your track and airspeed to get the missing bits, namely the true heading to fly and the resulting groundspeed which you use for flight planning. Then apply the magnetic variation and compass deviation to get the proper heading to fly.

CR Series

These were invented by Ray Lahr and marketed by Jeppesen. They are circular, with no sliding scale, and are based on trigonometry (they are easier to work with one hand, but be aware that, as the angle of drift increases, there's a small angular correction to be applied on top).

Below is a picture of how you would work out the PLOG above.

The cross between the 10 and 20 under the centre of the instrument is the wind velocity (180/15). Its position to the right of the main line going towards TC (True Course) means the wind is coming from the right, and the crosswind component is 2 kts. Looking across from the 20 on the outside scale (bottom right), you will see that 1.3° is the correction to be applied to obtain the heading (the white arrow above the letters TAS must be opposite the aircraft's TAS for this to be correct).

The tailwind component is 15 kts, which should be added to obtain a groundspeed of 105 kts.

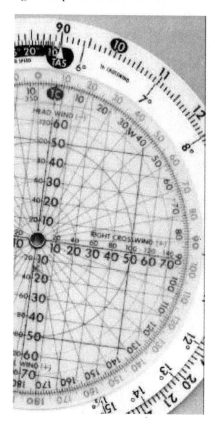

The very small ones have some functions left out, which are unimportant to most aircraft anyway, to pack everything else in, but don't get one too small, because your eyes won't see the print so well at night (see the *Human Factors* chapter).

Slide Rule

On the other side of both types, there is a circular slide rule, with the 60 point on the inner scale conveniently marked to make speed and time calculations easier:

It can be positioned against fuel quantity or distance on the outer scale to read time on the inner scale. As with any slide rule, you need the approximate answer to your problem first, as a protection against gross error and to give you an idea where to put the decimal point (if you were wondering how it works, you are adding indices, which is also where logarithms come from, but that is outside the scope of this book).

The most common problems concern time and distance. Just move the inner scale until the 60 point is opposite the TAS or fuel consumption, for example. Read the time on the inner scale against distance on the outer scale, or fuel if you are checking how much is being used. In the above picture, the *speed triangle* (60) is opposite 120 (knots or gallons) on the outer scale, which means it will take 6 minutes to go 12 nautical miles, or 6.5 to use 13 gallons, and so on. Always reduce hours (and proportions thereof) to minutes for simplicity.

To multiply normally (e.g. 2 x 4), place 10 on the inner scale against one number on the outer scale, and read the answer on the outer scale opposite the other, which will be on the inner scale.

Square roots can be found easily, too (useful for finding VHF ranges). Find the number you want the square root of on the outer scale, then rotate the inner one until the number opposite 10 is the same as the one against your original number. For example, 400 will have 2 opposite, as well as against 10 (figure out where the decimal point should go mentally).

To find TAS, line up the temperature against the pressure altitude in a window in the rotating slide rule (it may be labelled True Air Speed), then read the TAS on the outer scale against the RAS. Don't forget to allow for compressibility at speeds over 300 kts.

Conversions are done by lining up arrows on both scales representing the commodities concerned. For fuel weights, you will need the *specific gravity*, which is 1 for water, and used as a common denominator. It will vary from place to place, but that in the Flight Manual is the one to use. For example, if the s.g. of fuel is taken as .8, how much does 1 gallon weigh? The answer is 8 lbs (water would weigh 10). Alternatively, how many litres do you need to get from the fuel guy if you can carry 2600 kg and the s.g. is 8.2? Try 3170.

In the picture above, you will see that the arrow labelled km on the outer scale is opposite the one marked statute (miles) on the inner scale. All you do is read of the direct equivalent on each scale – here, 112.5 km is equal to 70 statute miles.

Note: Electronics are all very well, but batteries run out and electrics reserve the right to go wrong at the drop of a hat, as any avionics technician will tell you. My recommendation, at least in flight, is to use an E6B or a CR, because they are easier to work without getting your head stuck in the cockpit. The alleged accuracy you get with electronic computers is not worth the bother (and the expense), since you won't be able to read the instruments that closely anyway. *There is absolutely nothing wrong in non-ATP pilots who don't fly high speed aircraft using circular slide rules*, despite what other books may say.

Weight & Balance

This must follow the Flight Manual, to ensure an aircraft is safely loaded. As the Flight Manual forms part of the Certificate of Airworthiness (or Permit to Fly), if its conditions are not met, any insurance is invalid.

There are two aspects to Loading, the weights themselves and their distribution, and you sometimes get some nasty surprises—fuel in wings means unusually shaped fuel tanks, so you won't get a straight line variation; every fuel load will have a different figure, principally because the fuel tanks have a C of G system all of their own, running separately from the aircraft (even in small helicopters, like the Bell 206 or 407).

Every aircraft has a *Maximum Takeoff Weight*, which is the maximum with which you may get airborne (although you may not always be able to use it – see below). Any aircraft will fly overweight to a certain extent, if only because there's a tolerance range in the performance figures– ferry flights frequently do so, with the extra weight being fuel,

but having the physical ability doesn't mean that you should. You will at some stage be under some pressure to take an extra bit of baggage or top up with that bit of fuel that will save you making a stop en route, but consider the implications. Firstly, any C of A (and hence insurance) will be invalid if you don't fly the aircraft within the limits of the flight manual, and, secondly, you will be leaving yourself nothing in hand for turbulence and the like, which will increase your weight artificially. The designer will have allowed for 60-degree turns all the way up to MAUW, but not heavier than that. Even worse, your engine-out capabilities will be less than expected, particularly with autorotation.

There are very few light aircraft that will allow you to fill all the seats and cargo holds and still take full fuel. The effects of overloading include reduced acceleration capabilities (leading to longer takeoff and landing distances), decreased climb capability (watch for those obstacles), reduced range, ceiling, manoeuvrability, braking and margins, to mention but a few.

Here are some of the most common weights you will encounter:

- The *Empty Weight* is that in the *Weight and Centre of Gravity Schedule* (in the Flight Manual), which is established by actual weighing before the machine is used. It is the weight of the empty aircraft, plus unuseable fluids, and any fixed equipment.

- The *Basic Empty Weight* is the empty weight, above, plus operating fluids (fuel, full oil).

- The *Maximum Takeoff Weight*, which is simply the Basic Empty Weight plus the payload (passengers, cargo), which will not necessarily coincide with the full maximum, for performance reasons. The conditions at your *destination*, for example (it may be hotter and higher) may mean taking off lighter so you can land safely. Any weight less than the full maximum due to performance factors is known as the *Restricted (or Regulated) Takeoff Weight* (RTOW) and is actually the starting point for calculating payload (see below).

A couple of things to bear in mind are that the Basic Weight and the payload will not change during flight, but the fuel load will. Another is that, in small aircraft, passengers and cargo should be weighed separately.

Distribution

Incorrect loading naturally affects aircraft performance, and will possibly prevent the thing from even getting airborne. A Centre of Gravity too far forward will make it more difficult to raise the nose on take-off (or landing), possibly overstress the nosewheel as a result, and make the flight less economical by excessive use of trim tabs, which causes more drag. There are certain advantages to having the C of G towards the rear (by making the tailplane contribute to total lift, or at least not detract from it, which also reduces the power required and hence fuel used), but too much will make the aircraft less stable, more fatiguing to fly and cause similar drag and nosewheel problems (but in reverse) as excessive forward C of G. Also, if

you don't have the elevator movement to get yourself out of a stall, you could end up in a flat spin you can't get out of.

In a helicopter, if the C of G is too far aft or forward of its ideal position, there is a danger of running out of cyclic control in the opposite direction – one too far forward, for example, will mean you will not be able to pull the cyclic back far enough to cope with certain stages of flight (as fuel is consumed, for example, when the C of G generally moves forward), as a lot of its range will be taken up with the unusual attitude, although a forward position is needed to counteract flapback. Not being able to flare in an autorotation could well ruin your day (in fact, if your engine fails and you don't have enough cyclic movement to counteract the nose down tendency, the airflow will meet the disc edge-on and not go up through it, so you will not enter autorotation, and the RPM will decrease even more – ouch!). As well, *lateral* C of G may be affected with some loads, such as when hoisting.

The *reference datum* is an imaginary point from which all calculations start and where some C of G ranges are expressed (for example, 106" aft of datum). Mostly, it is at, or slightly forward of, the nose, but can be at the rotor mast of a helicopter.

The *arm* is the distance from the reference datum to the area in question, such as a passenger seat or the fuel tank. It may be measured in Imperial or Metric units, and you *must* use the same ones (the word *station* may also be used). To get the C of G of an aircraft, you multiply the weight of each item in it by the arm to get the *moment*, or the amount of leverage that item contributes.

The aircraft itself will have an arm and a moment from when it was last weighed, and this is where you start. You can find it in the *weight and balance schedule* (usually in the flight manual), and it may be varied if you add or take off various items of equipment, such as the hook or hoist in a helicopter.

Because you might end up using very long numbers, sometimes you use a *moment index*, the result of dividing the moment by 1,000 to make the figures more manageable. Here is a simplified typical calculation for a Bell 206 helicopter (the principles are the same for larger machines):

Item	Wt	Arm	Moment
Aircraft	1881	116.5	219137
Front pax	185	65	13000
Rear Pax	185	104	19240
Baggage	50	147.50	7375
Fuel	310	110.7	34273
Total	2611	112.22	293025

The total C of G for takeoff is 112.22, obtained by dividing the total moment figure (293025) by the total weight (2611). This particular machine's fuel has a *variable* CG range, meaning that it has one all on its own (that is, the arm will change with the weight), so the figure of 110.7 will change with the amount, for which check the flight manual.

The procedure is therefore to multiply the weights by the arms to get the moments, and divide the total moments by the total weights to get the C of G. Then you refer to

the flight manual to see if the figure fits into the authorised range. Look for a graph like this:

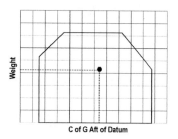

C of G Aft of Datum

Simply take the all-up weight you end up with, and the final C of G, and line them up horizontally and vertically. If they are inside the envelope, you are OK, but don't forget you have to land again! Your C of G may well be fine for takeoff, but check again after the fuel has been used!

Lateral C of G works the same way, except the figures are smaller and easier to work with. In fact, they may even be zeroed if the items are in the centre, as the fuel tank might be.

Items left of the centreline have a negative sign, and those on the right are positive, so the lateral moments for the front doors on a 206 would be –12 for the left and +12 for the right. Here's an example lateral C of G for the 206:

Item	Wt	Arm	Moment
Aircraft	1881	.41	773
Pilot	185	14	2590
Front pax	185	-11	-2035
Left Rear Pax	185	-16	-2960
Centre Rear Pax	185	0	0
Right Rear Pax	185	16	2960
Baggage	50	147.50	7375
Fuel	310	0	0
Total	3166	2.75	8703

Again, there will be a chart in the Flight Manual to show you where your plot lies. Fortunately, most of the time it is something that only helicopter pilots need to bother about, which is why it's mentioned.

Flight Manuals often have helpful charts with precalculated moment figures for fuel and baggage (the arm figures will be excluded). They are quite simple to use, except that the exam will require you to interpolate here and there. However, you should watch for special conditions, as with any chart, especially for maximum weights in particular locations. There may also be a plan view of the aircraft with the arms displayed next to the locations they refer to.

Performance

The take-off and landing phases are the most critical, demanding the highest skills from crews and placing the most strain on the aircraft. Because of this, strict regulations govern the information used for calculating take-off or landing performance. Of course, in the old days (say during the war, or when the trains ran on time), having enough engines to lift the load was all that mattered and no priority was given to reserves of power and the like. Now it's different, and performance requirements will be worked out before a C of A is issued, over a wide range of conditions. They are subsequently incorporated in the Flight Manual, which actually forms part of the C of A. In addition, the ANO requires you to ensure that your aircraft has adequate performance for any proposed flight.

It is your responsibility to decide whether or not a safe takeoff (and landing) can be made under the prevailing conditions. This means, in particular, that, although the crosswind on a particular runway may be within limits for your aircraft, you can choose another one if you are not happy.

Individual aircraft of a given species will vary in performance due to such variables as the age of the airframe and engines, the standard of maintenance, or the skill and experience of the crews. What you can do on one day under a given set of circumstances may well be impossible another time. The original testing, of course, is done with new aircraft and highly experienced pilots. "Performance" is therefore based on average values.

There are fudge factors applied to unfactored figures to produce *net performance* (and *gross performance* when they're not), which are meant to offset the effects of tired engines or variations in pilot skill. Occasionally, performance data in a flight manual will already be factored, but you will have to check the small print on the chart, in case they surprise you. Figures and graphs are based on standard conditions which allow for fixed reductions in pressure and temperature with height. As we all know, the real world isn't like that, so these assumptions may not always be true and due allowance must therefore be made for them (if your aircraft is performing sluggishly, you may find it's not the machine, but the conditions it has to work under that are at fault).

All the relevant data will be in the graphs, but some groups have no information at all in some areas. For instance, an aircraft may be assumed to have all engines working until above 200 feet, under which height there is no data for landing or take-off (which is why the take-off minima should not be below this, because you must be visual to avoid any obstacles should an engine fail). Sometimes, there can be no specific provision for engine failure at all.

Whatever you're flying, you will find the data needed to check your performance in the Flight Manual, which will have a supplement if your aircraft is foreign made, or you are using non-standard equipment— these override any information in the standard manuals. General principles concerning distances for take-off and landing are similar for aeroplanes and helicopters; for example, take-off distances for both will increase by 10% for each 1,000-foot increase in Pressure Altitude.

Accuracy with charts is essential – very often you have to interpolate between figures, and it's a good idea to get used to paralleling lines between the several graphs that may be on one chart.

Study the examples carefully and *always* read the conditions on which the chart is based – helicopter ones, for example, often need the generator switched off.

Factors affecting performance are:

Density Altitude

This is the altitude at which the ISA density is the same as that of the air in question or, in other words, your real altitude resulting from the effects of height, temperature, pressure and humidity, all of which can make the air thinner and which are mentioned below. The details will be in the Flight Manual, although humidity is usually ignored in the average performance chart, because it has more to do with engine power than aerodynamic efficiency, and high air density and humidity do not often go hand in hand. However, if the air is humid, say after a good shower, you would be wise to be careful.

Anyhow, the idea is that the more the density of the air decreases for any reason, the higher your aircraft thinks it is. If you look at the lift formula, you will see that the lift from a wing or thrust from a propeller is directly dependent on air density, as is drag, of course. The effects are as valid at sea level as they are in mountainous areas when temperatures are high – for example, 90° (F) at sea level is really 1900' as far as your machine is concerned. In extreme circumstances, you may have to restrict your operations to early morning or late afternoon.

Here is a handy chart:

°F/C	60/15.6	70/21.1	80/26.7
1,000'	1300	2000	2700
2000'	2350	3100	3800
3000'	3600	4300	5000
4000'	4650	5600	6300
5000'	6350	6900	7600
6000'	7400	8100	8800
7000'	8600	9300	1,0000
8000'	9700	10400	11100
9000'	11,000	11600	12400
1,0000'	12250	13000	13600
11,000'	13600	14300	15000
12000'	14750	15400	16000

It shows that, at 6,000 feet and 21°C, for example, you should enter performance charts at 8100 feet.

If you want to work it out for yourself, try this formula:

$$DA = 145,366[1 - (X^{0.235})]$$

where X is the station pressure in inches divided by the temperature in *Rankin* degrees, which are found by adding 459.69 to Fahrenheit totals.

Altitude
Air density drops off by .002 lbs per cubic foot (i.e. 2 ½ %) for every 1000 feet in the lower layers of the atmosphere.

Humidity
Adding water *vapour* to air makes it less dense because the molecular weight is lower (dry air is 29 –water vapour is 18). On cold days, humidity is less of a problem simply because cold air holds less vapour. A relative humidity of 90% at 70°F means twice as much than at 50°F.

Temperature
As heat expands air, it becomes thinner. Thinner air is less dense (*Boyles Law*). On the surface, an increase in temperature will decrease density and increase volume, with pressure remaining constant. At altitude, however, pressure reduces more than temperature does, and will produce an apparent contradiction, where temperature will decrease from the expansion.

Pressure
Air density reduces with atmospheric pressure (Charles Law). When you compress air, its density increases.

Runway length
Getting the wheels off the runway is only part of the story. You must also clear an imaginary screen (usually 35 or 50 feet) at the end of the TODA (TORA + Clearway). The distance for this is *Take-off Distance Required,* which must not be over TODA.

The *Stopway* may be added to the TORA to form the *Emergency Distance Available* (EDA), or the ground run distance available to abort a take-off and come to rest safely. EDA is sometimes also referred to as the *Emergency Distance* or *Accelerate-Stop Distance.* The greater the EDA, the higher the speed you can accelerate to before the point at which you must decide to stop or go when an engine fails.

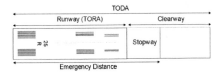

The *Landing Distance Available* must similarly not be less than the *Landing Distance Required.*

Altitude and Temperature
The higher you are, the less dense the air and the less the ability of the wings (rotating or otherwise) and engines to "bite" into it, thus requiring more power and longer take-off runs to get airborne. Humidity has a similar effect.

TODR will increase by 10% for each 1,000' increase in aerodrome altitude and 10% per 10°C increase in temperature (factor by 1.1). LDR will increase by 5% for each 1,000' increase in PA and 10°C increase in temperature (factor by 1.05).

Aircraft weight

Greater mass means slower acceleration or deceleration and longer distances. TODR will increase by 20% for each 10% increase in weight and LDR 10% per 10% increase in weight (factor by 1.2 and 1.1).

Some manuals give take-off and landing weights that should not be exceeded at specific combinations of altitude and temperature, so that climb performance is not compromised. These are known as *WAT limits (Weight, Altitude* and *Temperature)*.

Runway slope

Going uphill when taking off will delay acceleration and increase the distance required. The converse is true of downhill slopes and a rule of thumb is that TODR will increase 10% for each 2% of uphill slope, and vice versa (factor both by 1.1). When landing, an uphill slope aids stopping, thereby reducing LDR. Any gains from landing upslope or taking off downslope should not be made use of but accepted as a bonus (that is, don't use them as part of your planning).

Surface winds

Headwinds will reduce the distances required and improve the flight path after take-off. Tailwinds have reverse effects and crosswinds may even exceed the ability of the tyres to grip the runway. Aside from the handling problem, crosswinds may also increase the TODR if you need to use the brakes to keep straight. Forecast winds should be factored by 50% for a headwind and 150% for a tailwind. TODR and LDR will

increase by 20% for each tailwind component of 10% of the lift-off and landing speed (factor by 1.2).

The flight manual will state maximum crosswinds for your machine (try the *limitations* section). A useful guide (for American machines, anyway) is that the maximum crosswind will be about 20% of V_{so} (see below). When finding the angle between the wind and the runway, remember that runway headings are magnetic and forecast winds are true.

You can also use the crosswind chart to find a *limiting wind*, or the maximum you can accept from any given angle. Just draw a line upwards from the maximum speed you can accept, and stop when you reach the line representing the wind direction. The curved line at that point (or its interpolation) is the maximum windspeed you can take.

Surface

Performance information is based on a dry, hard surface. A "contaminated" runway has standing water or slush more than 3mm thick, or snow and ice anywhere along the takeoff run or accelerate-stop surface. However, your flight manual may have different ideas.

The most important factors are loss of friction when decelerating, and displacement of (and impingement drag when accelerating through) whatever is on it, so it may be difficult to steer, and take-off and accelerate-stop distances may increase due to slower acceleration, as will landing distance because of poor braking action and *aquaplaning* (see *Hydroplaning*, below), which is a condition where the built-up pressure of liquid under the tyres at a certain speed will equal the weight of the aircraft.

When taxying

On the ground, you may need slower taxying speeds and higher power settings to allow for reduction in braking performance and the increase in drag from snow, slush or standing water, so watch your jet blast or propeller slipstream doesn't blow anything into nearby aircraft.

Try not to collect snow and slush on the airframe, don't taxi directly behind other aircraft, and take account of banks of cleared snow and their proximity to wing- and propeller-tips or engine pods. Delay flap selection to minimise the danger of damage, or getting slush on their retraction mechanisms.

Hydroplaning

This occurs when liquid on the runway tends to creep under the tyres. Higher speeds will lift them completely, leaving them in contact with fluid alone, with the consequent loss of traction, so there may be a period during which, if one of your engines stops on take-off, you will be unable to either continue or stop within the remaining runway length, and go water-skiing merrily off the end (actually, you're more likely to go off the side, so choosing a longer runway won't necessarily help). The duration of this risk period is variable, but will vary according to your weight, the water depth, tyre pressure and speed.

Dynamic hydroplaning is the basic sort, arising from standing water. *Viscous hydroplaning* involves a thin layer of liquid on a slippery surface, such as the traces of rubber left on the landing area of a runway (one reason why it's dangerous to drive after a rain shower in Summer).

Reverted Rubber Hydroplaning happens when a locked tyre generates enough heat from friction to boil the water on the surface and cause the resulting steam to stop the tyre touching the runway. The heat causes the rubber to revert to its basic chemical properties.

A rough speed at which aquaplaning can occur is about 9 times the square root of your tyre pressures, 100 pounds per square inch therefore giving you about 90 kts (7.7 times if the tyre isn't rotating)—if this is higher than your expected take-off speed you're naturally safer than otherwise. The point to note is that if you start aquaplaning above the critical speed (for example, when landing), you can expect the process to continue below it, that is, you will slide around to well below the speed you would have expected it to start if you were taking off.

Under-inflating tyres doesn't help— each 2 or 3 lbs below proper pressure will lower the aquaplaning speed by 1 knot, so be careful if you've descended rapidly from a colder altitude.

Grass

For dry short grass (under 5"), the **TODR** will increase by 20%, a factor of 1.2. When it's wet, 25%—a factor of 1.25. For dry, long grass (5-10"), TODR will increase by 25%, and 30% when wet (it's not recommended that you operate when the grass is over 10" high).

For dry short grass (under 5 inches), the **LDR** will increase by 20%, a factor of 1.2. When it's wet, 30%—a factor of 1.3. For dry, long grass (5-10 inches), LDR will increase by 30%, and 40% when wet. For other soft ground or snow, the increase will be in the order of 25% or more for take-off and landing.

Obstacles (The Climb)

The best Rate of Climb speed is obtained when there is the greatest difference between the power required for level flight and that available from the engines. In turboprops, this will coincide with the speed that gives the best lift/drag ratio, since power output is relatively constant. Turbojets, however, produce more engine power with speed, which is enough to overcome the extra drag, so the maximum differential between power required and available happens at a higher speed. There will be performance tables to find time and fuel required for climbs. Remember that headwind and tailwinds will change the distance figures. To cope with this, work out the groundspeed with no wind and apply the corrections then. You can use the whizzwheel to find out the distance and time.

The Cruise

That part of the trip from the top of the climb (TOC) to the top of the descent (TOD.

For most trips, fuel management revolves around getting the maximum range for a given amount of fuel or, looked at another way, how little you can get away with on a fixed distance. However,

occasionally you must hold, and the question of how long you can stay airborne arises, namely endurance.

Speed

Peculiar to landing is speed—a higher one than specified naturally requires a longer distance, not only for slowing down, but the FAA have also determined that being 5 knots too fast over the threshold is the equivalent of being 50 feet too high.

Miscellaneous

Low tyre pressures increase distances required.

V-Speeds

Significant aircraft speeds:

Speed	Meaning
V_{NE}	Never Exceed speed, around 90% of Dive Speed (V_D). On a turbine, the equivalent is V_{MO} or M_{MO}.
V_{NO}	Normal Operations
V_{TOS}	In a single, a target speed to be achieved when becoming airborne. It allows safe control with a margin over stalling speed of at least 20%.
V_A	Manoeuvring speed. The maximum speed you can make abrupt, full scale deflections of the controls without damage (about twice V_S). Not always the best speed for entering turbulence (see below). Valid only at gross weight. A 20% decrease in weight needs a 10% reduction in manoeuvring speed.
V_F	Flap Speed
V_{FE}	Max speed flaps extended
V_{FO}	Max flap operating speed
V_H	Max level speed at max continuous power

Speed	Meaning
V$_{LE}$	Max speed, gear extended
V$_{LO}$	Max gear operating speed
V$_{LOF}$	Lift off speed
V$_{MCA}$	Minimum Control Speed, for control in the air, with critical engine out and the other at takeoff thrust.
V$_{S}$	Stall speed, or min steady controllable speed in flight.
V$_{SL}$	As above, for a specific configuration
V$_{SI}$	Stall speed, clean, power off
V$_{SO}$	Stall Speed in landing config.
V$_{2}$	Takeoff Safety Speed, or minimum safe flying speed if you lose an engine after takeoff, to be achieved before screen height (35'). Must be at least 20% more than stall speed and 10% above V$_{MC}$. Weight is the main factor. Should be calculated for each landing for go-arounds.
V$_{X}$	Best angle of climb
V$_{Y}$	Best rate of climb
V$_{YSE}$	Best s/e rate of climb

Charts

These may look complicated, but once you've figured out the way to enter a performance chart, it becomes considerably easier. For exam purposes, the best tip is to read the small print around the graph itself, as this is where you will find the conditions on which the chart is based, such as "generator off", or "ant-icing on".

Sometimes, the "chart" is not a chart at all, but a series of tabulated figures that require interpolation.

Height/Velocity Curve

Otherwise known as the *Dead Man's Curve*, this is a chart for helicopters that compares speeds against heights to give you an idea of where not to be if you want to make a successful recovery from an engine failure, that is, you don't want to be at high altitudes with low speeds, or low ones with high speeds, so the best place to be is in the gap between the shaded areas (see overleaf):

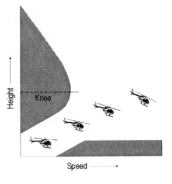

In other words, the graph shows combinations of speed and height that the average pilot would find it impossible to make a safe emergency landing from. To give the average pilot more of a chance, a one-second delay is factored in for minimum skill levels.

A couple of points to note first of all: one is that you should check to see if the chart is in the *Limitations* or *Performance* section of the Flight Manual. If it's in the latter, its requirements are *recommended*, not mandatory. In some circumstances, it is more dangerous to try to avoid the curve, especially if you might only be in it for a few seconds (as when getting out of a confined area, for example).

Anyhow, the vertical shaded area in the diagram above is called the *low speed* section, which is actually split in two parts at the knee of the curve, although it's never shown (the knee is the furthest point at which the curve extends). The lower portion is for takeoff power (no intervention), and the upper portion is for level flight (cruise power, hand not necessarily near the collective, so one second allowed for intervention time), and the whole area will expand with Density Altitude. On approach, you have your hands on the controls and are using less than cruise power, so the same figures don't work.

The other shaded area is called the *high speed section*, and the clear area between them is the *takeoff corridor*. Takeoffs and performance calculations should definitely take account of the curve, which is constructed at maximum weight, with no wind at a density altitude of at least 7,000 feet. Those of lesser quality must be verified (that is, flown) by the relevant Authority.

A Question

1. If you plan to fly for an hour and a half, and your machine uses 10 gals/hr, and you must land with 10 gals on board, what is your minimum fuel before takeoff?

An Answer

2. 25 gals.

Flight Operations

Fuelling

Jet and piston fuels mix differently with contaminants (particularly water), which is due to variations in their specific gravities and temperature. The specific gravity of water, for example, is so close to Avtur that it can take up to 4 hours for it to settle out, whereas the same process may take as little as half an hour with Avgas. As a result, there is always water suspended in jet fuel, which must be kept within strict limits, hence two filtration stages, for solids and water. The latter doesn't burn, of course, and can freeze, but it's the fungi that gather round the interface between it and the fuel that is the real problem – it turns into a dark-coloured slime which clings to tank walls and supporting structures, which not only alters the fuel chemically, but will block filters as well. Not much water is required for this – trace elements are enough, although, in reduced temperatures, dissolved water will escape as free water, and look like fog. Aviation fuel is "clean" if a one-quart sample is clear of sediment when viewed through a clean, dry, clear glass container, and looks clear and bright.

Note: It has been found that when visible water is present in jet fuel containing anti-icing additive, the additive will separate from the fuel and be attracted to the water. After a certain amount, thought to be about 15%, the density of the new liquid changes so much that it is not identified as water, and will therefore pass through water filters, and will not be detected by water finding paste. Where the ratio becomes 50%, as much as 10% of whatever is going through the filter could actually be water, which is very likely to get to the engine, since the filters on the airframe itself are not as restrictive.

Aircraft parked overnight should ideally have tanks completely filled to stop condensation.

An unofficial, but excellent (if not better) substitute for water paste or detectors when using jet fuel is food colouring, which you can at least get in the local grocery store, even if you're in a remote place. All you need is one drop – if there is no water, it will disperse evenly over the surface. If there is water, the food colouring will go directly to the water droplets, which will be more visible anyway from the colour.

Each day before flying, and when the fuel is settled, carry out a water check in aircraft and containers (but see below, for drums). Collect samples in a transparent container and check for sediment, free water or cloudiness—if there is only one liquid, ensure it is not all water. The instructions for using water detectors are displayed on the containers, but water-finding paste will not detect suspended water, and is as an additional test, not a replacement for a proper inspection.

Naturally, only competent and authorised personnel should operate fuelling equipment, who must also be fully briefed by their Company. In practice, of course, refuellers know very well what they're doing, but you should still be in full communication with them. In general, the following precautions should be taken:

- Documentation must reflect the fuel's origins and its correct handling.

- Vehicles used for transportation must be roadworthy and regularly inspected.

- Fire extinguishing equipment must be available and crews familiar with its use.

- Maintain a clear exit path for removal of equipment in emergency.

- The aircraft, fuelling vehicle, hose nozzle, filters or anything else through which fuel passes should be electrically bonded *before* the fuel cap is removed.

- Don't refuel within 100 feet of radar equipment that is operating. Only essential switches should be operated, with radio silence observed during fuelling.

- Avoid fuelling during electrical storms, and don't use bulbs or electronic flash equipment within the fuelling zone. Non-essential engines should not be run, but if any already running are stopped, they should not be restarted until fuel has ceased flowing and there is no risk of igniting vapours.

- Brakes or chocks should be applied, but some places require brakes off when near fixed installations.

- Take out rescue and survival equipment so if the thing blows up you have something to hand.

Most important is daily checking, before flying. Spilt fuel should be neutralised - move the aircraft or wait for it to evaporate before starting engines again.

Fuel can burn you. High vapour concentrations will irritate the eyes, nose, throat and lungs and may cause anasthaesia, headaches, dizziness and other central nervous system problems. Ingestion (like when siphoning) may cause bronchopneumonia or similar nasties, including *leukemia* and *death*. If you get it on your clothes, ground yourself before removing any and rinse them in clean water. Fuel spills on the ground must be covered with dirt as quickly as possible.

Otherwise, everybody not involved in the process should keep clear—at least 50m away, but for exceptions see later.

Fuel density changes with temperature -on a hot day, you won't get as much in, and will get less endurance. So, the colder the temperature, the heavier the fuel. In general, you can take avgas as weighing 6 lbs per US gal, Jet A at 6.8 lbs and Jet B at 6.5.

Bird Hazards

Prevention is better than cure, and you may like to avoid birds as much as possible. Notifications of permanent or seasonal concentrations of birds are sometimes in NOTAMs. Otherwise, keep away from bird sanctuaries or other areas where they may be expected, such as along shorelines or rivers in Autumn or Spring—migrating birds use line features for navigation as well, but they don't necessarily keep 300m to the right. Gulls seem to be struck the most often, and they hang around the seaside or rubbish tips.

Most birdstrikes happen between July-October, during daylight hours, hitting mainly engines and windshields. Noticeably fewer occur at height, so try to fly as high as possible, certainly above 1500 or even 3000 ft (40% of strikes occur on the ground, or during takeoff and landing. 15% occur up to about 100' agl. The highest so far hit a DC-8 at FL 390). Also, the lower you go, the slower you should be. Avoid high speed descent and approach—half the speed means a quarter of the impact energy. A short delay on the approach could mean the clearance of a group of birds, as they do move in waves. Groups of birds will usually break away downwards from anything hazardous, so try to fly upwards if possible. You could also use landing lights to make yourself more visible, especially where two are flashing alternately. Avoid freshly ploughed or harvested fields, and beware of updraughts in mountains areas, where birds will be trying to get some free lift. Birds are most active at dawn and dusk.

The impact force from a bird increases with the square of the speed—at 110 kts, the impact from a 1 lb bird can exceed 1200 lbs sq/inch (the force is actually determined by the square of your speed multiplied by the mass of the bird). The problem is that, below a certain weight of aircraft, the windshield will only be designed to keep out rain and insects. However, a hot windshield is more pliable and less easy to shatter—some aircraft need these on for take-off and landing, but if there is nothing in the flight manual about the optimum warmup time, use 15 minutes.

Overheating is as bad as underheating, so be wary if your aircraft has been left in the sun a long time. If you get a birdstrike, stop and inspect the damage immediately. If you can't, make sure you have controllability before trying to land again—fly the aircraft first.

Ditching

Ditching is a deliberate act, rather than an uncontrolled impact, although the terms are often used synonymously. A successful one depends on sea conditions, wind, type of aircraft and your skill, but it's the after effects, like survival and rescue that appear to cause the problems (88% of controlled ditchings happen without too many injuries, but over 50% of survivors die before help arrives).

Of course, the best way out of a ditching is not to get into one, but you can't always avoid flying over water. The next best thing is to prepare as much as possible beforehand, and make sure that the equipment you need is readily available, and not stuck in the baggage compartment where no-one can reach it. Have you *really* got enough fuel for the trip? Did you top up the oil or check the weather?

Once under way, flying higher helps in two ways, by giving you that little extra time to reach land, and to allow you to brief and prepare the passengers better. Maintaining a constant listening watch helps somebody know your position, as does filing a flight plan before going.

Sea Movement

It's a good idea to have a basic knowledge, as getting the heading right may well mean the difference between survival and disaster.

Whereas waves arise from local winds, swells (which relate to larger bodies of water), rely on more distant and substantial disturbances. They move primarily up and down, and only give the illusion of movement, as the sea does not actually move much horizontally. This is more dominant than anything caused by the wind, so it doesn't depend on wind direction, although secondary swells may well do. It's extremely dangerous to land into wind without regard to sea conditions; the swell *must* be taken into consideration, although it could assume less importance if the wind is very strong.

The vast majority of swells are lower than 12-15 feet, and the *swell face* is the side facing you, whereas the *backside* is away from you. This seems to apply regardless of the direction of swell movement.

The Procedure

You will need to transmit all your MAYDAY calls and squawks (7700) while still airborne, as well as turning on your ELT, or SARBE. If time permits, warn the passengers to don their lifejackets (without inflating them, or the liferafts) and tighten seat belts, remove any headsets, stow any loose items (dentures, etc.) and pair off for mutual support, being ready to operate any emergency equipment that may be to hand (they should have been briefed on this before departure).

One passenger should be the "dinghy monitor", that is, be responsible for the liferaft. If it's dark, turn on the cabin lights and ensure everyone braces before impact (the brace position helps to reduce the flailing of limbs, etc. as you hit the water, although its primary purpose is to stop people sliding underneath the lap strap; there are different ones for forward and aft seats).

If only one swell system exists, the problem is relatively simple—even if it's a high, fast one. Unfortunately, most cases involve two or more systems running in different directions, giving the sea a confused appearance. Always land either on the top, or on the backside of a swell in a trough (after the passage of a crest) as near as possible to any shipping, meaning you neither get the water suddenly falling away from you nor get swamped with water, and help is near.

Although you should normally land parallel to the primary swell, if the wind is strong, consider landing across if it helps minimise groundspeed (although in most cases drift caused by crosswind can be ignored, being only a secondary consideration to the forces contacted on touch-down). Thus, with a big swell, you should accept more crosswind to avoid landing directly into it. The simplest way of estimating the wind is to examine the wind streaks on the water which appear as long white streaks up- and downwind. Whichever way the foam appears to be sliding backwards is the wind direction (in other words, it's the opposite of what you think), and the relative speed is determined from the activity of the streaks themselves. Shadows and whitecaps are signs of large seas, and if they're close together, the sea will be short and rough. Avoid these areas as far as possible—you only need about 500' or so to play with.

The behaviour of the aircraft on making contact with the water will vary according to the state of the sea; the more confused and heavy the swell, the greater the deceleration forces and risks of breaking up (helicopters with a high C of G, such as the Puma, will tip over very easily, and need a sea anchor to keep them stable – in fact, the chances of any helicopter turning upside down are quite high). Landing is less hazardous in a helicopter because you can minimise forward speed. In fact, if you are intentionally ditching, you should come to a hover above the water first, then throw out the kit and the passengers. Having moved away from them, settle on the surface. If you can't do that, a zero speed landing should be aimed for, which means a steep flare a little higher and sooner than normal – any fore and aft movement on landing may cause rocking. Level off higher, as well.

You need to protect your thumbs throughout the whole process, as undoing a seat belt is a lot more difficult without them. Another tip is to reduce the length of your neck by hunching your head into your shoulders, like a turtle. Be particularly aware that anything happening to the blades will be transmitted through the controls, and may well be painful, or worse, if you get the cyclic in your stomach. At some stage you will be able to do nothing further with the controls, so be prepared to take place your limbs so that they do not flail about.

Keep the knees together, and prepare to use the hand near the exit to get out with, and the other to release the seat belt, but not until the machine is completely under water and has preferably stopped moving. This is to ensure you keep the same relative position to the chosen exit.

It will also provide extra leverage if you have to push against anything.

Once on the water, hold the machine upright and level using all the cyclic control there is, and use the rotor brake (if you've got one). Then let the aircraft sink. Rolling to ensure that the advancing blade is aft of the fuselage is one consideration, but this will increase the chances of disorientation, although it does ensure that the engine or transmission moves away from the cabin if it breaks free, due to gyroscopic precession. The way out of a submerged cabin is to place a hand on an open window or door, and follow your hand out, so you have a better idea of which way is up. Otherwise, instruct passengers not to leave until everything has quietened down. When you do, take the flotation and survival gear, but keep everyone together (remember that even seat cushions float). Attach the raft to the aircraft until you need to inflate it, as it will sail away downwind quite easily.

Splash, use flares or mirrors to attract attention, but let the rescuers come to you. Don't leave the security of the raft or aircraft unless you're actually being rescued as the downwash or wind will blow them away from you.

Keep moving—don't attempt to swim unless land is less than a mile or so away, but DON'T DRINK SEAWATER – it absorbs liquid and body fluids are used to try and get rid of it, so it gets you twice. Cold makes you give up, so try and keep a positive mental attitude. Except in mid-ocean, SAR will be operational very soon after the distress call, so

switch on the SARBE or ELT as soon as convenient, which will also assist a SAR satellite to get a fix on you. Try not to point the aerial directly at rescue aircraft as this may put them in a null zone.

Don't worry if the rescue helicopter disappears for ten minutes after finding you. It will be making an automatic letdown to your exact position after locating your overhead at height. This is where the temptation to use speech is very strong, but should be resisted because this is when the homing signal from the ELT/SARBE is most needed. Speech should only be used as a last resort as, not only will it wear your batteries down, but also take priority over the homing signal used to fix your position. If you feel the need to do something, fire off a few mini-flares instead. Or scream.

Finally, once in the winch strop, don't grasp the hook, because of the possibility of shocks from static electricity.

Equipment
This needs to be for aviation use.

Rafts
Aviation liferafts are designed to vent to atmosphere in case of a problem, rather than into the liferaft itself, as is the case with marine ones (they could inflate in the cabin).

As it will float before it's fully inflated, tie it to the airframe (unless it's actually sinking), or a person, before inflating (in fact, it should be tied to at least one person as much as possible). Do this downwind, so it doesn't get damaged against the aircraft. To turn it upright in the water, get downwind, and place the cylinder, which is heavy, towards you. This weight, plus the wind, will help it to flip over. Once inside the raft, protect yourself as much as possible with the canopy, and get the sea anchor out. Buoyancy chambers should be firm, but not rock hard.

Lifejackets

An unconscious person needs 35 lbs of buoyancy to keep afloat, so make sure they are so capable, especially taking a fair bit of wear and tear. Automatically inflated types activate when a soluble tablet gets wet, which is no good in a water-filled cabin, as you will be unlikely to get out of the cabin entrance. Purloining them from airlines is also not a good idea, as they use one-shot jackets. The reason CO_2 is used to inflate them is that it doesn't burn.

Immersion Suits

Immersion suits are useful, but they are not necessarily to keep you warm long-term, that is, to delay hypothermia, although that is part of their function - a good majority of deaths with a suit on occur well within any time needed for hypothermia to even set in. The real danger is inside the first two or three minutes, from cold shock response, which will reduce your capacity to hold your breath, and possibly set off hyperventilation, aside from contracting blood vessels and raising the blood pressure. At temperatures between 5-10°, the average capability for holding the breath reduces to about 10 seconds, if at all.

From 3-15 minutes, the problem appears to be keeping the airways clear – it can be quite frustrating trying to breathe while you're

continually being splashed. It's not till 30 minutes have passed in average conditions that hypothermia starts to rear its head, and if you're not wearing a lifejacket, it will reduce your ability to use your arms to swim. Even the method of taking you out of the water can be dangerous if it causes the blood to pool away from the cardiovascular system – whilst in the water, its pressure against your body helps return blood from the lower limbs back to the heart – this support is removed once you are out.

Night Flying

Night flying can be pleasant—there's less traffic, you tend not to go in bad weather and the air is denser, so the engine and flying controls are more responsive (if the controls become heavier than normal, your instructor has his hands on as well!).

Searching for an overdue aircraft in low light conditions causes lots of problems, and route planning should take account of this. Otherwise, it's much the same as for day, though there are some aspects that demand some thought. Plot your route on the chart in the normal way, but navigate with electronic aids or features that are prominent at night, such as town lighting, lighted masts or chimneys, large stretches of water (big black holes), aerodromes, highways, etc.

Apart from reducing visibility, rain on the windscreen is a particular threat when fixing your position by a single light source. When little or no light is on the surface and a prominent one comes into view, it may seem that the light is above the horizon, which could lead you to pitch into a steep attitude in keeping with the resulting false horizon. Sometimes the effect is not much more than an uncomfortable climbing sensation even when you're straight and level, but an obscured windscreen could make objects appear lower than they really are.

This will be more apparent with high intensity runway lighting, which may also give you the same effect that actors have on stage, where they can't see the audience through the bright lighting. The lack of normal contrast will also upset your altitude perception, making you feel further away and higher than you are. As a result, on a final approach you could find yourself too low and fast. The solution is to use every piece of sensory information you can, including landing lights and instruments (look ahead and slightly to the side of the light beam).

Problems will arise if several of the above factors affect you at once, especially if the landing point is sloping—this is where frequent cross-checking of altimeters is important. The illusions you might get with sloping ground include:

Problem	Illusion	Risk
Downslope	Too low	High approach
Upslope	Too high	Low approach
Rain	Closer	Low approach
Narrow	Too high	Low approach
Wide	Too low	High approach & flare
Bright lts	Too low	High approach

The trick with landing is to get to the point where you think the wheels are going to touch the ground – then go down another 30 feet.

Helicopter landing sites should be checked out in daylight on the same day as they are to be used at night. Preflight checks should allow for night flying—carry a torch, and 2 landing lights are preferred. Permission to enter the rotor disc is given by flashing landing lights. Hovertaxi higher and slower than by day, making no sideways or backwards movements. Great care should be exercised in pointing the Schermuly flares to a safe place at all times (which is admittedly a bit difficult when they're fitted and the fuelling truck pulls up right alongside them). The flares should not be armed at this stage, but at the holding point immediately before take-off and disarmed at the same place after final approach. They should also be disarmed after reaching cruising altitude.

The maximum useful height for discharging a flare is approximately 1800'. Its burn time is 80 seconds, during which time it will fall about 1500'. Therefore, having established autorotation after an engine failure at night, the first flare should be discharged immediately, or on passing through 1800 feet, whichever is later. Don't do it before this, as they will be useless. If possible, the second should be discharged between 800-1000' agl.

In night autorotations, use a constant attitude, at whatever speed is comfortable, which keeps the beam from the landing light in the same position on the ground, or it will shine up into the air when you flare, from which position it's no good to you at all.

Air Law

This chapter covers enough of the UK Air Navigation Order and Rules of the Air for exam purposes, as well as JAR rules, which were formally adopted on 1 July, 1999. For more details about how law works with respect to aviation in UK, refer to my other book, *The ANO (& Rules of the Air) in Plain English*.

The ANO itself consists of *Articles* and *Schedules*, the latter being amplifications of the former, so where an Article would require an aircraft to carry markings, the related Schedule would spell out how they are to be made and positioned. All UK aircraft are subject to the ANO and Rules anywhere at any time, although in a foreign state, you must obey its laws and report the occasions you do so to the CAA.

JAR, by the way, stands for *Joint Airworthiness Requirements*, which works on the premise that aviation is the same in most civilised countries, and can be standardised to a certain extent. Essentially, certain European countries have agreed upon common procedures to help with importing and exporting aircraft, type certification and maintenance between them, based on existing European regulations and FARs (from the FAA in the USA), where acceptable. In fact, the maintenance side of JAR, 145, is directly drawn from FAR Parts 43 and 145. Naturally, there's a committee somewhere that jollies things along, which is somewhere in Holland, and the bottom line is that it is easier to use foreign aircraft

However, licensing has not been left out. JAR-FCL covers using licences and ratings between member states with the minimum of formality. JAR-FCL 1 refers to aeroplanes, FCL 2 to helicopters, FCL 3 to medicals, FCL 4 to flight engineers (one day), and FCL 5 to balloon and glider licences, all based on ICAO Annex 1, with suitable amendments.

As well as using standard aviation phrases as shorthand, and a few acronyms, it is assumed that all activities will be carried out safely, with any equipment required being approved and serviceable, unless allowed under an MEL or with CAA permission, to save unnecessary repetition (and a lot of typing), and that the word 'approved' means 'approved by the CAA'. The word 'Convention' means the *Chicago Convention*, as amended. Also, remaining engines (when one fails) will operate within maximum continuous power conditions in the C of A. Acronyms include (but are not limited to):

AIP	Aeronautical Information Publication
AGL	Above Ground Level
AMSL	Above Mean Sea Level
APU	Auxiliary Power Unit
ATC	Air Traffic Control
C of A	Certificate of Airworthiness
C of R	Certificate of Registration
CPL	Commercial Pilot's Licence
CRM	Crew Resource Management
DOCs	Direct Operating Costs
MTBF	Mean Time Between Failure
PF	Pilot Flying
PNF	Pilot Not Flying
PPL	Private Pilot's Licence
SOPs	Standard Operating Procedures
TCA	Terminal Control Area

Tips For Reading Legal Stuff

Simply take the confusing passage and split it up into separate lines based on where commas appear in the text.

Medicals

Flight Crew Licences, other than radio licences, are not valid without a medical certificate, which is renewed from time to time.

You may not act as flight crew if you know or suspect that your physical or mental condition renders you temporarily or permanently unfit to do so. In other words, you may not exercise any licence privileges once you become aware of a decrease in your medical fitness that makes you unable to exercise them safely.

Medicals are only valid if you meet the initial issuing requirements. A Board of Inquiry or insurance company may interpret the words "medically fit" a little differently than you think if you fly with a cold or under the influence of alcohol. In any case, you should talk to a medical examiner as soon as possible in the case of:

- admission to a hospital or clinic for over 12 hours

- surgery or other invasive procedures

- regular use of medication

- regular use of correcting lenses

In addition, you should inform the CAA in writing of significant personal injuries involving your capacity to act as a member of a flight crew, or illness that lasts for more than 21 days (after the 21st day), or pregnancy. In these cases, your medical is suspended.

JAR

Only issued for professional and private licences for aeroplanes and helicopters. For Flight Engineers, etc., refer to *UK National*, below.

Class 1

Although intended for professional licences, a PPL holder may hold one at any time. The initial issue must be done at the CAA Medical Centre at Gatport Airwick.

Class 2

Required for the PPL. The initial issue can be done by any aviation medical examiner.

If your national licence does not meet Class 2 standards, you can still exercise the privileges of that licence.

UK National

Class 1
For Flight Engineers, Navigators, and CPL (Airships).

Class 2
CPL (Balloons) on public transport, with the initial one at Gatwick.

Class 3
For CPL (Balloons) on aerial work, and PPL (Airships) and (SLMG). Initials and renewals can be done by any aviation examiner.

Declaration of Health
An alternative to the Class 3 for private microlights, powered parachutes, gyroplanes or balloons. It is signed by the PPL holder and countersigned by any GP or AME.

Licences & Ratings

You must have a valid licence and rating that complies with JAR-FCL to fly as crew in a JAA-registered aircraft. The licence itself may have either been issued by a JAA member state, or come from another ICAO state, and been validated.

Having said that, within the UK, the Channel Islands, and the Isle of Man, you can operate radios without a licence as the pilot of a UK registered glider not flying for public transport or aerial work and not talking to ATC, or if you are being

trained in a UK registered aircraft for flight crew duties, when authorised by a licence holder. The messages must only be for instruction, or the safety or navigation of the aircraft, on frequencies exceeding 60 MHz assigned by the CAA, which are automatically maintained, and if the transmitter has external switches.

For other licences, where an aircraft is being towed, no-one else must be on board unless they are authorised or are a trainee. Ex-military pilots (within the last 6 months) or flight engineers/navigators upgrading to pilot or adding types may also fly without licences under supervision. Military pilots may fly without licences in the course of their duties.

You can also be a PIC for renewal of a pilot's licence or inclusion or variation of ratings if you are at least 16 with a medical certificate (and comply with its conditions), under the supervision of an instructor and nobody else is carried. It must also be a non-public transport or aerial work flight, although your instructor can get paid. On single-pilot aircraft, duals must be fitted for instructors.

You can act as PIC of a helicopter or gyroplane at night if you do not have an IR and have not carried out at least 5 takeoffs and landings at night in the last 13 months, under an instructor, with nobody else on board, and not engaged on public transport or aerial work, except paying the instructor. The same applies to balloons if you have not done 5 flights of over 5 minutes.

You don't need a licence to fly a glider unless you are operating the radio or you are engaged on public transport or aerial work, other than aerial work consisting of instructing or testing in a club aircraft and the people concerned are members.

You need permission from a Contracting State if you fly in or over it in a UK registered aircraft and your (UK) licence does not meet the full specification. You also need permission from the CAA if your foreign licence is not up to scratch and you fly in or over the UK.

Refer also to Article 26 (3) about medical fitness and pregnancy.

Licences are granted under Part A of Schedule 8, assuming you are over the minimum age, are fit to hold them and are qualified, having provided suitable evidence. They are not valid without the signature of the holder in ink.

A JAR licence is considered valid for ANO purposes. A non-JAR foreign licence does not allow you to be paid for public transport or aerial work, fly IFR, or give instruction.

A JAR licence may be transferred to another state if you are employed there or have established normal residence, which means you live there for at least 185 days in the year, with enough personal or occupational ties to confirm it.

JAR licences are valid for up to 5 years.

You can't hold a licence just by itself – there must be some sort of type rating on it. Licences must bear a valid certificate of revalidation for the rating under Section 2 of Part C of Schedule 8.

If you fail a test required for articles 23, 24, or 25, you cannot use the

privileges of the relevant rating, even if there is time left on the original.

Private Pilot Licence

Unless stated otherwise, references to training and experience concern appropriate classes of aircraft.

You must be over 17 to hold one, although you can act as PIC from 16 years of age, under an instructor, in UK airspace. Dual instruction before 14 does not count

There is no maximum validity for UK PPLs - JAR ones last for 5 years.

Privileges

JAR

You can fly as PIC or P2 of an aeroplane, helicopter or gyroplane of any type or class (just type for helicopters) within a rating in the licence, but not for public transport or aerial work, except instructing, if you are an instructor, or testing, which must be in a machine owned, or operated under a flying club where you and the victim (sorry, student) are members. Towing gliders and dropping parachutists may also be done. No pay may be received, except by instructors in microlights or self-launching motor gliders, under the conditions above.

Unless you have an IR or IMC rating, you may not be PIC outside controlled airspace when the vis is below 3 km, under Special VFR in a control zone below 10 km, except on an approved route, or out of sight of the surface. You may not be PIC or P2 IFR in Class D or E airspace, and you need an IR for PIC in IFR in Class A, B or C airspace.

A night rating or qualification is required to be PIC at night.

NPPL

You can be PIC of any simple single-engined aeroplane, microlight or SLMG specified, or otherwise within any rating on the licence. Outside the UK, however, you must comply with Art 21(10)(a), with permission from any contracting states, except for an SLMG if you have an appropriate medical and SLMG rating.

The only aerial work you can do is towing another aeroplane or glider in a club aircraft, if you and any passengers are club members, or any other approved organization.

You cannot fly a simple single-engined aeroplane outside controlled airspace in under 5 km visibility, or a microlight or SLMG under 3 km. Neither may you fly Special VFR in a control zone under 10 km, out of sight of the surface, at night, or when needing to be IFR.

Unless you have appropriate differences training, the simple aeroplane may not have a retractable or tricycle undercarriage, a tailwheel, a supercharger or turbocharger, a variable pitch propeller, a cabin pressure system or a max continuous cruising speed over 140 kts. The training must be recorded in your log book.

Microlights may not have 3-axis controls or flexwing controls if your previous experience has been with the other type, unless you have difference training and record it in your logbook.

Medical

You need either a Class 1 or 2 medical certificate.

Recency

You must maintain competency under JAR-FCL 1.025 p1-A-5.

To carry passengers, you must do, in the previous 90 days, 3 takeoffs and landings (included in 3 circuits for helicopters) as sole manipulator in the same type or class (for a NPPL, the maximum number of passengers is 4). With no IR, at least 1 takeoff and landing must be at night. For gyroplanes, you need at least 5 take offs and landings when the centre of the sun was at least 12° below the horizon in the last 13 months, or 5 5-minute flights in free balloons, for balloons.

Validity

For single-engined piston aeroplanes and touring motor gliders, you must pass a proficiency check in the three months before the rating expires. Alternatively, you can do 12 hours (6 hours PIC) and 12 takeoffs and landings or at least one hour with an instructor (or other test) in the previous 12 months.

Experience

At least 45 hours, to include 25 hours dual and 10 hours supervised solo (in the State that is going to issue the licence), with at least 5 hours solo cross country with a flight of at least 270 km (150 nm) having two full stop landings at different airfields.

For helicopters, the 25 hours dual must include 5 hours' instruments, and the cross-country flight must be at least 185 km (100 nm)

Flight Radiotelephony Operator

You must be at least 16. The maximum validity is 10 years, if standalone, or the validity of an associated licence. It s normally issued for VHF only, unless you pass a separate HF exam, or the CPL or ATPL radio nav exams.

A JAR PPL will be tested as art of the skills test for the licence. The FRTOL is a UK national licence.

The licence is needed unless you are a student obtaining an additional rating under a flying instructor, so if you are solo, you may not use the radios if you don't have one.

It is also illegal if you (as PIC) don't have an R/T licence to allow anyone with one to operate the radios.

Privileges

You can use the radios in any aircraft if the frequency is maintained automatically, using external switching devices.

Foreign Licences

Licences from Contracting States (i.e. ICAO) are valid for private purposes on G-registered aircraft automatically, as long as the original licence remains valid and the pilot does not get paid, except for instrument or instructor ratings.

Otherwise, the CAA can validate foreign non-JAR licences at any time, including those under the law of an EEA state (see also Annex 1 to the Chicago Convention). Such licences must be based on equivalent licences under Art 22. The Commission must be asked for an opinion on licence equivalency within three weeks of receipt by the CAA of all necessary information, or inform you of extra requirements within three months, if it doesn't actually grant the validation.

However, it must, within three months of receipt of the necessary information for the application, either issue the certificate of validation or inform you of any additional requirements or tests.

IMC Rating

Along with some other special ratings, this is restricted to the sovereign airspace of the issuing state, in this case the UK. Its use elsewhere needs the permission of the relevant state, as it is not a full instrument rating.

It is for aeroplanes only, and enables a PPL or BCPL (aeroplanes) holder to be PIC inside the bad visibility restrictions of both licences, in the UK, except on a special VFR flight in a control zone with visibility less than 3 km, or taking off or landing where the flight visibility below cloud is less than 1800 metres.

It allows private flight under IFR outside controlled airspace, and in Class D, E and F airspace.

Night Rating

This allows a PPL or BCPL holder to be PIC at night. There is a special one for gyroplane PPLs.

Flight Instructor - Art 29

You may not instruct any one learning to fly for a licence or rating unless you are qualified as PIC on the aircraft concerned, and have an instructor rating.

With *restricted privileges*, the licence is restricted until you have done at least 100 hours flight instruction and supervised at least 25 solo flights, and have been recommended. That is, you can do (under supervision) instruction for PPLs, or those parts of integrated courses for single-engine aeroplanes, excluding approval of first solo flights and cross country flights, and night flying instruction.

You can do up to 5 hours in an approved simulator.

Recording of Flight Times

Normally, flight time to be credited towards a licence must be flown in the same category of aircraft as the licence (or rating) sought.

All solo, dual or PIC flight counts in full towards the requirements for a higher licence. 50% of P2 (co-pilot) time is counted, but JARs counts it in full on machines which require more than one pilot.

Personal Logs – Art 28

These must be maintained by every member of the Flight Crew of a UK registered aircraft and, regardless of registration, those qualifying for licence purposes. They may be computerised for commercial flights.

Aside from your name and address, log books should also have the date of the flight, name of PIC, type of aircraft, registration, crew position, places and times of departure and arrival, flight times, operational conditions, such as night or IFR, and details of any tests taken.

Flight time is the total time from when an aircraft first moves under its own power with the intention of taking off until it comes to rest after the flight, so it includes taxi time (for helicopters, the flight time stops when the rotors do). This is what goes in your log book. *Air time*, on the other hand, is between wheels or skids off and when they touch the Earth again. This is what goes in the Tech Log.

A logbook is your personal and private property, not having been issued to you under the ANO.

If you lose it, you will need to provide a Sworn Affidavit that details your flying hours to the best of your knowledge, if you intend to rely on them to get further licences and ratings.

Flight time records must be produced without undue delay upon a request from an Authorised Person. Student pilots must carry them on solo cross-country flights because they contain records of instructor authorisations.

Rules of the Air

Low Flying – Rule 5

You may not fly over any congested area of a city, town or settlement below a height that allows you to land clear and without danger to persons or property on the surface if a donkey stops (in a helicopter, you just need to alight without danger, not necessarily clear). If you're towing a banner, you must take that into consideration as well.

Alternatively, you can proceed at 1500 feet above the highest fixed object within 600 metres, whichever is higher, unless on a notified route, landing or taking off, on Special VFR or you are a police aircraft.

A helicopter must be able to alight clear in the *London Helicopter Zone.*

Unless you are a police aircraft, you may not fly over, or within 3000 feet of, any open air assembly of more than 1000 people witnessing or participating in any organised event, except with permission in writing from the CAA and the organisers, or below a height that allows you to land clear if an engine fails. Banners should not be dropped within 3000 feet. This doesn't count if you are taking part in an aircraft race or contest, or a flying display.

You may not fly closer than *500 feet to any person, vessel, vehicle or structure* (exam question) unless you are operating a police aircraft, or landing or taking off under normal aviation practice, hill soaring with a glider, working under an aerial application certificate, or picking up or dropping tow ropes, banners or similar articles at an aerodrome. However, be aware that this is under discussion.

Nothing in this rule stops you flying as necessary for saving life, or applies to any captive balloon or kite. Neither does it stop any aircraft flying under normal aviation practice, when taking off from, landing or practising approaches at, or checking navigational aids or procedures at, a Government aerodrome, one owned or managed by the CAA or a licensed aerodrome, in the customarily used airspace.

Lights & Signals – Rule 8
If more than one light is needed to comply with the Rules of the Air, only one should be visible at a time. Where a light must show through specified angles horizontally, the light should be visible from 90° above and below. Lights showing in all directions must be visible from any point horizontally and vertically.

Display of lights by aircraft – Rule 9
By night, only lights in the Rules may be shown, to avoid confusion. By day, anti-col lights must be used in flight (a red one with engines running when stationary).

At night, a flying machine on a UK aerodrome must display normal lights, or any in Rule 11(2)(c), unless stationary on the apron or maintenance park. Lights may be reduced or switched off if they affect crew duties or dazzle people.

Navigation lights are set up so that only one can be seen by another aircraft at any time. Anticollision lights are seen from all directions.

Failure of lights – Rule 10
On the ground, if a light cannot be immediately repaired or replaced, the aircraft may not depart. In flight, you must land as soon as you can safely do so, unless otherwise authorised by ATC. If an anti-col light fails by day, you may continue, but the light must be fixed at the earliest possible opportunity.

Flying machines – Rule 11
At night, a UK registered flying machine over 5700 kg, or has a first

type on or after 1st April 1988, must show a steady green light of at least five candela to starboard through 110° from ahead horizontally, with a steady red on the other side, and a steady white of at least three candela through 70° each side of dead astern horizontally. If more than 2 metres from the wing tip, another may be put at the tip.

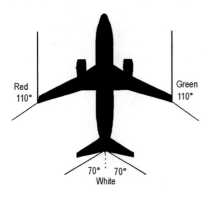

Red 110°

Green 110°

70° 70°
White

Note that the boundaries re parallel, so only one light can be seen.

Where the type certificate is dated before then, and the aircraft is below 5700 kg, as above, plus an anti-collision light or flashing white light of at least twenty candela showing in all directions, flashing alternately to the main lights. Otherwise any of the above may be used.

Gliders – Rule 12
At night, either a steady red light of at least five candela, showing in all directions, or lights under Rule 11(2) and (3).

Free balloons – Rule 13
At night, a steady red light of at least five candela showing in all directions, suspended between 5-10 metres below the basket, or, if there isn't one, below the lowest part.

Captive balloons and kites – Rule 14
When over 60 metres agl, two steady lights, one white, 4 metres above the red, both at least five candela and showing in all directions, with the white light between 5 and 10 metres below the basket or the lowest part of the balloon or kite. On the mooring cable, at intervals less than 300 metres from the lights above, groups of two lights as above, with one below the cloud base if the bottom one is hidden by cloud.

On the surface, a group of 3 flashing lights arranged horizontally in an approximately equilateral triangle, around the item to be marked, with sides at least 25 metres. One side must be approximately at right angles to the horizontal projection of the cable and delimited by two red lights - the third must be green.

A captive balloon by day over 60 metres agl must have tubular streamers on its mooring cable every 200 metres, measured from the basket or lowest part. They must be at least 40 centimetres in diameter and 2 metres long, marked with alternate bands of red and white 50 centimetres wide. A kite in similar circumstances can have the streamers above, or every 100 metres measured from the lowest part, at least 80 centimetres long and 30 centimetres wide, marked with alternate bands of red and white 10 centimetres wide.

Airships – Rule 15
At night, or while picking up moorings (even if under command, that is, able to execute manoeuvres),

a steady white light of at least five candela showing 110° either side of dead ahead, a steady green light of at least five candela showing to starboard through 110° from dead ahead horizontally, a steady red on the other side, and a steady white of at least five candela showing through 70° either side of dead astern horizontally, plus an anti-collision light. If not under command, has stopped engines or is being towed, the white lights mentioned above, two red ones, each at least 5 candela and showing in all directions below the control car, with one at least 4 metres above the other and at least 8 metres below the car, and, if making way (but not otherwise), the green and red lights referred to above.

While moored to a mast in the UK by night, white light of at least five candela showing in all directions at or near the rear. Otherwise, a white light of at least 5 candela showing through 110° either side of dead ahead horizontally, and a white one of at least 5 candela through 70° either side of dead astern.

By day, not under command, with engines stopped, or being towed, two black balls below the control car, one at least 4 metres above the other and at least 8 below the car.

Right of Way – Rule 17

To make life easier, there are rules about how aircraft should be flown when mixing with other traffic. Even when you have right of way, you must take any necessary action to avoid collision (in other words, even with clearance, commanders are responsible for avoiding collisions).

Another aircraft with an emergency gains priority over you. If you must give way, you must not pass over or under, or cross ahead of, the other aircraft unless you are far enough away not to create a risk of collision. Aircraft with the right of way must maintain course and speed.

Aircraft may not be flown so close to others that a danger of collision is created, and may not fly in formation unless the commanders aircraft have agreed amongst themselves.

A glider and whatever is towing it are regarded as one aircraft under the PIC of the towing machine.

Convergence

If steady relative bearing is kept between two aircraft at the same altitude, they will eventually collide.

When two aircraft are converging in this way, the one from the right has the right-of-way, except that:

- power-driven, heavier-than-air aircraft (that is, flying machines) give way to airships, gliders and balloons

- airships give way to gliders and balloons

- gliders give way to balloons

- power-driven aircraft give way to aircraft that are towing or carrying a slung load.

When two balloons are converging at different altitudes, the higher one must give way.

Aircraft with the right of way should maintain course and speed.

Approaching head-on
If there is a danger of collision, each must alter course to the right.

Overtaking
You are overtaking when you are approaching another aircraft from behind at less than 70° from the longitudinal axis, which means that, at night, you should not be able to see its navigation lights.

Aircraft being overtaken have right of way, and the overtaking aircraft, whether climbing, descending or in horizontal flight, must keep out of the way by altering course to the right (to the left on the ground, so you can be seen from the pilot's seat) until well past and clear, even if their relative positions change. Gliders in UK may go right or left.

Landing and take off
There must be no apparent risk of collision with any aircraft, person, vessel, vehicle or structure in your take-off or landing path.

You must take-off and land in the direction indicated by the ground signals, or into wind if there are none, unless good aviation practice demands otherwise.

A flying machine or glider must not land on a runway if it is not clear of other aircraft unless ATC say otherwise. When not using a runway, you must leave clear on your left any aircraft which have landed, are already landing or are about to take off. You must make any turns to the turn left after checking that you will not interfere with other traffic.

A flying machine about to take off must leave clear on its left any aircraft that have already taken off or are about to.

After landing, move clear of the landing area as soon as possible, unless told otherwise by ATC.

Order of landing
Except where ATC dictate otherwise, or in emergency, aircraft landing or on finals have right of way over others in flight or on the ground or water. Where several are involved in landing, the lowest has right of way, as long as it does not cut in front of another on finals, or overtake it. During such an emergency at night, even if you have permission to land, you may not do so until you get further clearance.

Near aerodromes
A flying machine, glider or airship near what the commander knows (or ought reasonably to know) to be an aerodrome, or moving on one, must, unless told otherwise by ATC, conform to the circuit or keep clear. All turns must be made to the left unless ground signals say otherwise.

Aircraft Manoeuvring on Water
Aircraft or vessels coming from the right have the right-of-way.

If you are approaching another aircraft or vessel head-on, you must alter heading to the right.

Aircraft or vessels being overtaken have the right of way, and overtakers must alter heading enough to keep well clear.

Right Hand Traffic - Rule 19
An aircraft in the UK in sight of the ground and following a road, railway, canal or coastline, or any other line

of landmarks, must keep them on its left, except under instructions from ATC inside controlled airspace.

Simulated Instrument Flight

When your vision is restricted by artificial means, the aircraft must have fully functioning duals and a safety pilot qualified on type, plus an observer, if necessary.

Aerodromes & Airports

An aerodrome is generally any place for landing aircraft that fits the official definition, which is, broadly, being set apart for the purpose, and includes any necessary buildings. An airport is an aerodrome with different documentation, so it has to meet a higher level of safety. It would typically be used for passenger-carrying commercial flights, for example.

Full details of handy places to land, including heliports (see below) are in guides like Pooley's or Bottlang's (a mention here is not necessarily a recommendation), although the official document is the *UK Air Pilot*.

An aerodrome or airport listed in it that does not need previous permission for use is for *public use*. Where permission is required, you either need to get it first, or just provide *prior notice*, so they can get the sheep off the runway.

Although not in the regulations, in general, it is still a good idea not to:

- walk, stand, drive, park anything, or cause an obstruction, without permission from operators/ATC.

- tow aircraft on active movement areas at night, without wingtip, tail and anti-collision lights, or light from the towing vehicle.

- park, or otherwise leave an aircraft, on an active manoeuvring area at night without wingtip, tail and anti-collision lights, or lanterns suspended from the wingtips, tail and nose.

- operate a vessel on, or cause obstruction on the surface of, a water area that must be kept clear, when ordered to leave, or not approach it, by ATC.

- remove, deface, extinguish or interfere with markers, markings, lights or signals for air navigation, without permission from the operator and ATC.

- display a marker, marking, sign, light or signal likely to be hazardous by causing glare or confusion with, or preventing clear sight of, another one.

- allow a bird or animal in your custody to be unrestrained inside an aerodrome or airport, except to control others.

- discharge a firearm within or into an aerodrome or airport without permission from the operator.

- display a marker, marking, light or signal that may cause a person to believe a place is an aerodrome when it is not.

The above are not exam questions.

Operations nearby

There must be no likelihood of collision with other aircraft or vehicles, and the aerodrome must be suitable for the intended operation. This means observing other traffic and conforming to or avoiding the traffic pattern.

Standard Traffic Circuit

Make all turns to the left, arriving overhead the field at 2000' in a descending turn, or at 45° to the downwind leg, at the height published in the AIP (all circuit details will be in there, too). Only join directly downwind if there is no conflict.

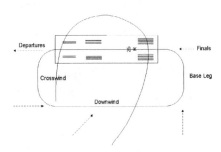

For right hand circuits, apply the opposite.

All the above, of course, can be varied by ATC at any time.

Helicopters

Hover taxying is movement in ground effect at speeds up to about 20 kts, but more likely the normal walking pace. The height may vary because of external loads, but the pace will be relatively slow.

Air Taxi is almost like flight, but below 100 feet, often used when ATC would like you to expedite your movement to help with traffic flow.

If you've got wheels, it is usual to taxi on the ground, but ATC aren't always aware that you have them.

Runways & Taxiways

Details are declared by the Airport Authority and published in the AIP, although they can be found in many other publications. This declared distance is either the *Take-off Run Available* (TORA) or *Landing Distance Available* (LDA). Any areas at the ends unsuitable to run on, but still clear of obstacles, are called *Clearways*, which, with the TORA, form the *Take-off Distance Available* (TODA), which should not be more than 1 ½ x TORA.

Part of the Clearway that can support an aircraft while stopping, although not under take-off conditions, is declared as *Stopway* which may be added to the TORA to form the *Emergency Distance Available* (EDA), and marked with yellow chevrons. This is the ground run distance available for an aircraft to abort a take-off and come to rest safely—the essential point to note is that Stopway is ground-based. EDA is sometimes also referred to as the *Emergency Distance* or *Accelerate-Stop Distance*. The greater the EDA, the higher the speed you can accelerate to before the point at which you must decide to stop or go when an engine fails.

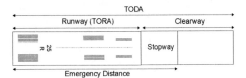

The end of the runway is called the *threshold*. Any obstacles interfering with the glideslope may need it to be *displaced* a certain distance, but the area behind it can still be used for taxying and takeoff runs, even if it cannot be for landing.

Displacement is marked by large yellow arrows pointing towards the new threshold. The threshold will be *relocated* if part of the runway is closed, and crosses will be used instead of arrows.

Runways are named after the direction they are facing in, without the last number.

For example, one facing West, or 270°, would be called *Runway 27*. In fact, the naming is to the nearest tenth degree, so one facing 067° is actually Runway 07. A T after the number (as in 07T) would be a True direction, as used in the NDA.

Parallel runways will also be known as *Left* or *Right*.

Where no runways are available, the takeoff and landing areas will be marked out with *pyramidal* or *conical* markers, painted orange and white for airports, or just orange for aerodromes.

Clearances to enter, land on, take off from, cross and back track on runways *must be read back*.

A *Long final* is a turn onto final approach more than 4 nm from touchdown.

Contamination
Whenever there is water on the runway, surface conditions are described (with its depth) as:

- *Damp* – there is a change of colour due to moisture.

- *Water patches* – significant patches of standing water.

- *Wet* – the surface is soaked, but there is no standing water.

- *Flooded* – there is extensive standing water visible.

Markings & Signals
For a 2500m landing distance, the distance from the threshold to the aiming point is 400m, with 6 pairs of markings on a 2500m runway.

Two or more white crosses (with arms at 45° to the center line) along a section or at both ends of a runway or taxiway mean that the section between them is unfit for aircraft movement:

The white T with a disc above (for airborne machines) and a single black ball suspended from a mast (for those on the ground), mean that the directions for takeoff and landing are not necessarily the same.

A mandatory instruction sign has white text on a red background.

Taxiways
Paved taxiways should have centre line markings, for continuous guidance between runways and aircraft stands.

Holding Points
A non-instrument runway will have a yellow single solid and a single dashed line across the taxiway (the

dashed line will be on the *runway* side). An instrument runway has a double set of each (see overleaf).

They indicate a point beyond which no part of an aircraft may project without prior permission from ATC.

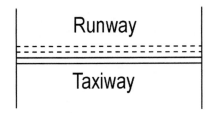

Identification Beacons
Military aerodromes have a flashing beacon at night coloured red, showing a two-letter Morse group.

Passenger Briefings

The regulations impose on you the responsibility for the safety and well-being of your passengers. The Common Law also imposes on you a duty of care to your neighbour, so it is a good idea to brief them before every flight, or at least take all reasonable steps to do so.

A lot depends on what your passengers are going to do at the destination – if you're going to shut down, then tell them to stay seated until everything stops (it helps to explain why you have to sit there for 2 minutes). If it involves a running disembarkation, one passenger should be briefed to operate the baggage door and do the unloading. Everyone else must leave the rotor disc area. Similar action must be taken with a running pickup.

Nobody should enter the area of ground covered by the main rotor disc of a helicopter without your permission (indicated by "thumbs up" during the day, or a flash of the landing light by night):

Movement in and out of this area should be to the front or at 45º to the longitudinal axis, ensuring that all movement is within your field of vision. Additionally, no movement should be allowed during startup or rundown (due to the dangers of blade sailing) and nobody should approach the rear of a helicopter AT ANY TIME. You can help by landing in such a way that passengers have no choice but to go forward, but watch the doors aren't forced against their stops if the wind is behind you.

Transistor radios, tape recorders and the like should not be operated in flight as they may interfere with navigation equipment. If you don't believe me, tune to an AM station, as used by ADF, on a cheap radio and switch on an even cheaper calculator nearby—you will find the radio is blanked out by white noise. In fact, the radiations from TVs and radios come within the VOR and ILS regions as well. Cellular phones are dodgy, too, but when you're up in the air, you also log on to more than one cell, which screws up the system, whereupon the FBI get upset because they can't find you (cell phones can be tracked). In any case, using mobile phones is against the ANO, aircraft radio licence and telephone licence, because they interfere with aircraft systems.

Anyway, as I said, you, as commander, are responsible for ensuring that all passengers are briefed, or have equipment demonstrated, as outlined below.

Pre-flight

Before take-off and landing (and whenever you deem it necessary, e.g. during turbulence), they also need to be told about the dangers involved in various aspects of aircraft operation, in particular the location of all exits and the use of safety equipment required to be carried, but also:

- Your authority as aircraft Commander.

- Methods of approaching the aircraft, in particular avoiding exhausts and tail rotors—if nearby aircraft have their engines running, it could mask the sound of a closer one. Pitot tubes are especially sensitive (and hot!). Children should be kept under strict control. Wait for signal from pilot. Used crouched position in pilot's view. Take off loose objects, clothing, hats, etc

- Dangerous Goods and hazardous items that must not be carried. No objects above shoulder height – carry equipment horizontally. Long

items should be dragged by one end. Do not throw cargo.

- Methods of opening and closing cabin doors (inside and outside) and their use as emergency exits. Not leaving seat belts outside. Where not to step and what to hold on to. Sharp objects must be handled carefully when working with float-equipped helicopters.

- Hazards of rotor blade sailing and walking uphill inside the rotor disc while rotors are running.

- When they can smoke (not when oxygen is in use!).

- Avoidance of flying when ill or drunk—not only is this dangerous to themselves, but if they are incapable next to an emergency exit, others could suffer too.

- How to use the seat belts and when they must be fastened.

- What not to touch in flight.

- Loose articles, stowage (tables, etc.) and dangers of throwing anything out of the windows or towards rotor blades.

- Use and location of safety equipment, including a practical demonstration (if you intend to reach a point more than thirty minutes away from the nearest land at overwater speed, you need to do this with the lifejacket, maybe in the terminal). When oxygen needs to be used in a hurry, adults should fit their masks before their children.

- The brace position (including rear-facing seats). If you ever have to give the order to adopt it, by the way, don't do it too early, otherwise the passengers will get fed up waiting for something to happen and sit up just at the point of impact.

- Landing areas should be clear.

- How long the flight will be, and how high you will be flying, what the weather will be like.

Preparation for Emergency Landing

Where time and circumstances permit, this must consist of instructions about safety belts or harnesses, seat backs and tables, carry-on baggage, safety features cards, brace position (when to assume, how long to remain), and life preservers.

Airworthiness & Maintenance

An aircraft is "airworthy" when it complies with the flight manual, any placards, and the ICAO Airworthiness Technical Manual.

As far as maintenance goes, it can be *Scheduled* or *Unscheduled*, which basically speak for themselves. Both are supposed to ensure that an aircraft is kept at an acceptable standard of airworthiness. Depending on the performance category and its maximum authorised weight, there will be different schemes covering this, but the nature of General Aviation means that aircraft are very often not seen by an engineer from one scheduled check to the next (but the pilot can do some elementary tasks – see below).

Types of check include 50-hour and 100-hour, which can be extended by 5 or 10%, respectively, for scheduling, but this should not be used as part of normal operations (lack of planning on your part doesn't justify an emergency on an engineer's part).

In between, there will also be times when components need to be changed, either on a planned or emergency basis.

The *Maintenance Schedule* contains the name and address of the owner or operator and notes the type of aircraft and equipment fitted.

It lays down the periods when every part of the machine will be inspected, together with the type and degree of the inspection, including periods of cleaning, lubricating and adjustment. They are written for each aircraft, and are subject to Transport Canada approval before moving to a new one.

After work is done, an *Aircraft Maintenance Engineer* (AME) signs a *Certificate of Maintenance Review*, which means that the work done meets the standards and the aircraft is released back into service. However, you are still responsible for ensuring that the aircraft is airworthy.

Not being an engineer, the only way you can find this out (aside from a thorough preflight) is to check the Technical Log before flight, in which you should find an alert card which shows when the next servicing is due. Simply subtract the current aircraft hours from that figure to find out how many hours' flying you can do before the next check.

After an abnormal occurrence (such as a lightning strike or a heavy landing), the aircraft must be inspected (and not flown until it has been done). If nothing has to be taken apart, the inspection can be done by the PIC, but I would suggest you need some technical qualifications to know that you don't need to take anything apart in the first place.

Dual Inspections

Required when engine or primary flight controls have been modified, repaired, replaced or disassembled.

Away from base, this may be carried out by a pilot qualified on type.

Elementary Work

This is technically a form of maintenance but, for ANO purposes, it means specific tasks not subject to Maintenance Review, which means you don't need an AME to do them.

A licensed pilot, who is also the owner or operator of an aircraft, may perform elementary work on it. Under normal circumstances, it is limited to aircraft under 2730 kg with a Private or Special Category C of A (see Regulation 16 of ANGRs). This might include changing spark plugs or tyres, batteries, or bulbs, although 17 are mentioned (the key is that no special tools are required, or that the structure is not affected).

Details of the work, that is:

- tyre and skid replacement

- elastic shock absorber cord unit replacement on landing gear

- safety wiring or split pin replacement (except on engines, flight control or rotor systems)

- simple patch repairs to fabric (not rotor blades), if they don't cover up structural damage or require removal of structural parts or control surfaces

- repairs to the cabin interior (including upholstery)

- repairs to non-structural fuselage parts, not requiring welding

- side window replacement, if the structure is not affected

- safety belt or harness replacement

- replacement of seats

- replacing bulbs, reflectors, glasses, lenses or lights

- replacement of cowlings

- replacing spark plugs

- replacing batteries

- replacement of wings and control surfaces that are designed to come off before and after flight

- replacement of rotor blades that are designed to come off

- replacement of generator and fan belts that are designed for removal

- VHF radios that are not combined with navaids

must be entered into an appropriate log, and certified. As said above, elementary work does not need a maintenance release.

Documents & Records

The Chicago Convention requires that these documents (or copies) *must* be carried on all flights – *JAR Ops 1.125* refers. Each is fully explained in its own section below:

* *Certificate of Airworthiness* or equivalent flight authority

* *Certificate of Registration*

* *Noise Certificate* (if applicable)

* *Radio Station Licence*

* *Air Operator Certificate*

* *Flight Crew Licences*, valid, with ratings, to include radio and medical certificates

* 3*rd* *Party Liability Certificate*

It's a good idea to carry these as well:

* *Certificate of Maintenance Review*, which must be valid before and during the flight. The PIC is responsible for ensuring this, and that the aircraft is airworthy.

* *Flight Manual*

* *Aircraft Weight Schedule*, which may be in the Flight Manual. It must be preserved up to 6 months after the next schedule is prepared (Art 18).

* *Technical Log* (but see below)

* *Route Guides and Charts*

* *Flight Plan*

* *Operations Manual and SOPs*, if operating under an AOC

Flight Authority (C of A)

All aircraft must have a document that says they are fit for flight.

Unless surrendered, suspended or cancelled, such a flight authority remains in force for the time or number of flights specified, or indefinitely, if the aircraft continues to meet the conditions under which the authority was issued.

Certificate of Airworthiness

Issued at the beginning of an aircraft's service, this remains valid by proper loading, obeying the flight manual, completing all scheduled maintenance, complying with all ADs, performing proper daily checks (as per the flight manual) and having all no-go defects rectified before flight. *It becomes invalid if the aircraft is repaired, modified, and/or any of its equipment or the airframe itself is overhauled* (Art 9). Otherwise, it is issued for a three-year period.

Aircraft flying in UK need a C of A from the State of registration, except (when only operating in UK) for kites, or balloons on private flights, privately operated gliders (that is, not on public transport or aerial work, except instruction or flight testing in club aircraft where the instructor and trainee are members), aircraft with a permit to fly (see below), or those working under A and B Conditions in Part A of Schedule 3, where it is assumed that the aircraft is flown for the purposes indicated (it also specifies the performance group for Article 36(1)).

A C of A remains in force for the period specified, or until the aircraft (or necessary equipment) is overhauled, repaired or modified, or equipment removed or replaced in an unapproved fashion, or the completion of mandatory servicing or modifications. If it has been endorsed by an ICAO state as failing to meet standards at certification, the aircraft concerned must have permission from relevant states to fly internationally.

Items to be checked regarding the certificate before flight include making sure the nationality and registration match, as does the type designation and serial number, the date of issue, the signature and seal.

In UK, there are different types, including *Transport or Aerial Work, Special Category* or *Private*.

Permit to Fly
These are required to fly in situations where the aircraft is still safe, but there is no C of A, or it has become invalid, such as when a maintenance check runs out away from base and you have to get the machine home.

It's more typically used for homebuilts or ex-military aircraft.

Certificate of Registration
All aircraft must be registered, either in their home, or a contracting or foreign state. There are exceptions for sales or test purposes (for which you need a *dealer certificate*), but these will not be for revenue flights.

An aircraft can be registered as:

- a *state aircraft*, i.e. a civil one owned and exclusively used by a government.

- a *commercial aircraft*, used for commercial air transport.

- a *private aircraft*.

In UK, the CAA is the relevant authority, and maintains the Register of Aircraft.

Generally, any citizen or properly incorporated body of the Commonwealth can be "qualified owners", and registration normally becomes void if an unqualified person acquires a legal interest or share. An 'interest' does not include one to which you are entitled only by being in a flying club, and the term 'registered owner' includes, for a deceased person, his legal personal representative, and the successor of a dissolved body corporate.

You must inform the CAA of any changes in the original particulars, the destruction of the aircraft, its permanent withdrawal from use, or termination of a demise charter. If you become the owner of a UK registered aircraft, you must inform the CAA in writing within 28 days.

Identity for Registration Purposes
Registration marks must be inscribed on a fireproof identification plate (metal is preferred), which must itself be in a prominent position near the main entrance (or conspicuously fixed to the exterior of the payload for an unmanned free balloon). Plates must have at least the nationality and registration marks.

When the plates are removed, the aircraft officially ceases to exist.

Noise Certificate
All aeroplanes require a noise certificate, except STOL aircraft (i.e. those that need a TODR less than 610 metres on a hard level runway at sea level in ISA at MAUW).

Radio Station Licence
This is automatically renewed each year, as long as the fee is paid.

Air Operator Certificate
Needed if you fly for public transport, that is, charge to carry people or freight. It is granted by the CAA. Government departments have to follow the rules, too, which is why the Police have AOCs so they can charge between forces.

A private aircraft will not need one as no charging is allowed.

Certificate of Maintenance Review
Needed by UK aircraft with Transport or Aerial Work certificates (of airworthiness), showing when the last review was done and when the next is due. The periods will be specified in the C of A.

Certificates of Maintenance Review must be issued in duplicate, only by holders of an AME licence (either under the ANO or a validated foreign one), or people authorised by the CAA for a particular case, who must verify that maintenance has been carried out under the schedule, that mandatory inspections and modifications have been completed as per the Certificate of Release to Service, defects in the tech log have been rectified or deferred, and Certificates of Release to Service have been issued.

One copy of the most recently issued certificate must be in the aircraft, except when taking off and landing at the same aerodrome when remaining in the UK, and the other kept by the operator somewhere safe for up to 2 years after its issue (unless Article 80 disagrees).

Flight Manual
Otherwise known as the *Pilot's Operating Handbook* (POH), this document is law with respect to operating your aircraft.

Extra equipment on your machine not already mentioned in it may have an added page or two in the back, called of a *supplement*, which may restrict or extend the limits imposed by the manufacturer.

Aircraft Weight Schedule
A weight schedule must be prepared for every flying machine and glider with a C of A, showing either the basic weight (e.g. empty with unusable fuel and oil, and equipment in the weight schedule) and the C of G for the basic weight, or approved alternatives.

Weight schedules must be kept for 6 months after the next weighing, but see Article 80.

Technical Log
A system for recording defects and maintenance between scheduled servicing, as well as information relevant to flight safety and maintenance. In other words, the formal communication between flight crews and engineering.

A tech log is required by every UK registered aircraft with a transport or aerial work category C of A, unless it is below 2730 kg MAUW and not operated by an AOC holder, where an approved alternative may be used. It must be filled in at the end of every flight, by the commander, specifically noting the takeoff and landing times, any known defects that affect airworthiness or safe operation (or *nil defects*) and anything else required by the CAA from time to time. Entries must be signed and dated. Those relating to defects must be readily identifiable. For a series of consecutive flights on the same day at the same aerodrome with the same commander (except spraying), the details may be filled in at the end of the last flight unless a defect must be reported.

Tech logs must be carried in the aircraft (see Article 76), with copies of the entries above kept on the ground, except for helicopters and aeroplanes weighing less than 2730 kg, where they may be carried on board in an approved container, if it is not reasonably practicable to do otherwise. They must be preserved for 2 years after the aircraft has been destroyed or permanently withdrawn from use, unless Article 80 disagrees, or the CAA varies it.

Route Guides & Charts

You need a route guide so you can get around the airways without messing things up for anybody else. Any used must be the current ones, as amended. Although the symbols and keys are generally printed on them, it is still a good idea to know at least the basic ones for use in flight in a hurry. Because they are included on the map, they have not been reproduced in this book.

Flight Plan

There are many reasons for filing flight plans – first of all, they help get you slotted into the system, even if it isn't quite the route you asked for. Next, they help with radio failures, as, once you're in the pipe, so to speak, everyone knows where you're supposed to be (more or less)

and can act accordingly. Then there are forced landings, where an educated guess may be made as to your position, followed by statistics, and, finally, because the law says you must, under certain circumstances (International flights, for example, always require a flight plan, as does IFR flight in controlled airspace).

You should always file a plan as far in advance as possible, but at least 60 minutes is preferable, with the ATSU at the departure aerodrome, but if there isn't one, you can do it by telephone or radio (in that order) to the unit designated to serve the aerodrome. If there is a delay of more than 30 minutes for a controlled flight, or an hour for an uncontrolled flight, you should either amend the flight plan, or cancel it and submit a new one.

If you land at an aerodrome other than the one in the flight plan, you must tell ATC within 30 minutes of the ETA at the planned destination.

Here are the details for each slot:

Aircraft ID

The call sign. In the absence of a company callsign, use the aircraft registration.

Flight Rules & Type of Flight

The former goes in Box 8. V=VFR, I=IFR, Y=IFR/VFR, Z=VFR/IFR (enter the changeover point in the route section). Where the change is composite, (e.g. VFR/IFR/VFR) the first takes precedence (Y).

Next is the *type* of flight. For the first character use C=Controlled VFR, D=Defence VFR, E=Defence Itinerary, F=Flight Itinerary. For the second, G=General Aviation,

S=Scheduled, N=Non-scheduled, M=Military, X=Other.

No & Type of Aircraft, Wake Turbulence Category

Box 9 is also in two parts. The first is the number of aircraft, which should be blank for VFR.

Equipment

Box 10, for comms, nav and transponder, in that order (COM, NAV, SSR). N=None or unserviceable. S=Standard, that is, VHF, ADF, VOR and ILS. C=LORAN, D=DME, F=ADF, G=GPS, H=HF RTF, I=INS, J=Data Link, K=MLS, L=ILS, M=OMEGA, O=VOR, R=RNP certified, T=TACAN, U=UHF, V=VHF, W=RVSM certified, X=MNPS certified, Y=CMNPS certified, Z=other.

For SSR, N=Nil, A=Mode A, C=Mode C, X=Mode S without ident and PA, P=Mode S with PA & no ident , I=Mode S with ident & no PA, S=Mode S with ident & PA.

For example, SD/C is commonly used for standard equipment with DMA and Mode C transponder.

Departure Aerodrome

In Box 13, use the ICAO code.

Departure Time

Anticipated time in hours and minutes UTC, over 60 minutes ahead.

Cruise Speed, Altitude, Route

Box 15 is for the flight planned TAS. N=Knots, M=Mach Number.

For cruising level, A=Altitude in hundreds of feet ASL (e.g. A050). F is for Flight Level.

Under *route*, include speed and altitude changes, airway numbers and waypoints on the route. DCT (*Direct*) is assumed unless they are included. IFR routes should be used when available, as per IFR charts.

Destination, EET, SAR, Alternate

In Box 16, use the ICAO code. The EET can include the number of days. Use your own SAR time, up to 24 hours after ETA.

Other Information

Use 0 if you have nothing else to add.

Supplemental Information

Fuel endurance in hours and minutes goes in Box 19. Place an X through the U and V if you do *not* have the VHF and UHF emergency frequencies (243 & 121.5 MHz).

Cross out the survival equipment you don't have, and add with whom the arrival report will be filed.

Finally, include your name and licence number, and the person to be notified if SAR is initiated.

NOTAM

A NOTice to AirMen is a temporary warning or notice relevant to flight, concerning changes to frequencies or serviceability of navaids, or hazards. They are in the list of items to be checked before flight and can be obtained by telephone, from an ATC office or over the Internet.

Class I Notams are distributed quickly over the AFTN (*Aeronautical Fixed Telecommunication Network* – the equivalent of telex for aviation organisations). Class II Notams are less urgent and are sent by post.

A Class 2 Notam may confirm a Class 1 Notam.

Notams do not amend the *Air Pilot* (see below), but they may affect the information it contains.

AIP

The *Aeronautical Information Publication* is a summary of the rules and regulations that affect aviation (similar documents are issued by all countries). As such, it is not the final authority for the rules you have to obey, and will not be produced in court, but the law that backs it up will. A clue as to what is or isn't supported by law is given by the use of the word "shall".

In the UK, it is a big blue book, otherwise known as the *UK Air Pilot*, and is split into eight sections:

- *AGA* – information on aerodromes, including safety altitudes, etc.

- *COM* – communications, radio frequencies, navaids, etc.

- *MET* – weather services, radio frequencies, met offices.

- *RAC* – rules of flight for the area, flight plans, reporting, etc.

- *FAL* – arrival, transit and departure procedures, customs, documentation, cargo.

- *SAR* – search and rescue organisation and procedures.

- *MAP* – maps and charts.

- *GEN* – general information unsuited to the above sections, such as time zones, registration marks, etc.

The above sections are the same from country to country so you can find the information easier.

When the ANO says that airspace is *notified* (see the *Glossary*), it means that its details are published in the AIP so you can take notice of them.

As with any publication of a like nature, it is amended regularly, and you should always make sure yours is up-to-date.

AIC

Aeronautical Information Circulars contain amendments to information in the AIP, but, again, they do not officially amend the publication (this is done by replacing complete pages occasionally). Pink AICs concern safety matters which should be brought to everyone's attention (they are Very Important). The others are coloured like this:

- White – Admin

- Yellow – Operational

- Mauve – Airspace restrictions

- Green – Maps and Charts

AICs have their own serial number, based on the year and number, such a 27/2003, but they will also be called something like *Pink 27* or *Yellow 42*, in brackets afterwards.

SUP

AIC SUPs contain temporary items of operational significance and comprehensive text and/or graphics (e.g.: major air exercises or aerodrome work) that make them awkward for Notams, although a Notam may be used to indicate changes to the period of validity (which should normally be in the SUP itself) or cancellation. They are issued every 28 days, and should be kept in the AIP as long as all or some of its contents remain valid.

Lights & Visual Signals

Acknowledge by rocking the wings or flashing the landing lights once.

Light or pyro	To aircraft in flight	To aircraft or vehicle on ground
Continuous red light	Give way to other aircraft, keep circling	Stop
Red pyro, or flare	Do not land; await permission	
Red flashes	Do not land; aerodrome not available	Move clear of landing area
Green flashes	Return, await permission to land	Move on manoeuvring area (and apron for aircraft)
Continuous green light	You may land	You may take off (not a vehicle)
Continuous green light, flashes or pyrotechnic		
White flashes	Land here after continuous green, then, after green flashes, proceed to the apron.	Return to starting point on aerodrome.

Visual Approach Systems

Runway lighting works backwards from the threshold for up to 3,000

feet for precision approach runways. It has a purpose other than to show you the way in at night – it's also meant to help you transition to the visual after emerging from the clouds during an approach. There are various designs for various purposes, shown in the AIP.

When approaching visually to a runway, it's often useful to have an aid to help get the glideslope right (lateral guidance is provided by the runway lights). Those described here use different coloured light patterns to show whether you are on a glideslope, too high or too low. They will be situated to the left of the runway threshold and visible up to about 5 nm by day and 20 nm or more by night. Their sphere of influence is ±10° of the extended centreline, up to 4 nm.

VASIS

The *Visual Approach Slope Indicator System* is a group of four lights (2-bar), which may be turned off if the weather is down to minimums so they don't confuse you.

The light bars are called *upwind* and *downwind*. There's a *middle* one for the 3-bar version - normal aircraft use the middle and downwind ones, and widebodies use the middle and upwind bars to get a glideslope of 3° (usually).

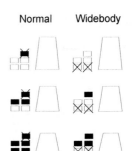

Normal   Widebody

Normal means an eye-to-wheel height of up to 25 feet (e.g. DC-8). Widebodies have up to 45 feet.

When you are on the glideslope, you should see red lights over white ones ("red on white, you're all right"). If you are too low, you will see red over red ("red on red, you're dead").

When the approach is correct, you will have safe clearance from obstructions within 6-9° either side of the centreline up to 4 nm out, with a safe wheel clearance over the threshold.

A *Tri-Color VASI* uses red, green and amber to indicate too low, on the glideslope and too high, respectively.

A *Pulsating VASI* (PVASI) uses a single light source to project a two-colour approach indication. When very low or very high, the light pulsates more in relation to your distance away, otherwise it is steady for on the glideslope (white) and just below (red).

PAPI

The *Precision Approach Path Indicator* does pretty much the same thing as VASIS, but with 4 lights in a row:

When on the correct slope, the two lights nearest the runway are red and the two furthest ones show white. Three whites and a red mean slightly high, and three reds and a white means slightly low. Four of each is way too much.

T-VASIS

This uses 10 lights, with 4 horizontal ones in the middle and the other 6 as 3 vertical groups above and below, which only appear when you are low or high. If you do things properly, you will arrive at the threshold at 45 feet. All lights are white, except where a *gross undershoot* is involved, when they turn red.

Lights above the horizontal 4 are the *fly down* lights, whereas those appearing below are the *fly up* ones.

Visual Flight Rules

Although the airspace you fly in comes in six varieties (see below), it is essentially split into two types, *controlled* or *uncontrolled*, although it's fair to say that, in UK, once you are above 3000 feet, most airspace is controlled in one form or another. As the names imply, in the first you do as you're told (by ATC), and in the second, you are responsible for the safe conduct of the flight, which means avoiding obstacles and other aircraft, which you can only do if you can see them.

The official definition of a flight under *VFR* is "one conducted under *Visual Flight Rules*", conveniently leaving out the bit that tells you what the Rules are (1500' ceiling with an associated ground visibility of 5 nm).

The *Visual Flight Rules* govern flight in *Visual Meteorological Conditions*

(VMC). When the weather gets so bad that you can't see where you are going, *Instrument Meteorological Conditions* (IMC) apply, and you must fly under *Instrument Flight Rules* (IFR), discussed below, although you can fly IFR at any time. The definition of IMC is actually a negative one, being "weather precluding flight in compliance with Visual Flight Rules". The point at which this happens depends on the type of airspace you are in.

Special VFR Flight

This is used when you can't comply with IFR in a control zone (in fact, in UK, there is no VFR at night, except Special VFR in a control zone). It's a legal technicality, used to allow VFR aircraft to go where the law says only IFR aircraft may fly. However, it is never volunteered - *you must request Special VFR*. ATC will provide separation between Special VFR and IFR flights.

Aside from requiring 10 km inflight visibility (as a PPL holder without any special ratings), you must be clear of cloud and in sight of the surface, and obtain clearance from ATC before doing so, which means you must have radios. You are therefore absolved from the 1500 foot rule (within 600 m – Rule 5), but not from being able to glide clear of a built-up area in an emergency.

If your radios stop working before entering the zone, you must remain clear of controlled airspace.

Helicopters should fly slowly enough to avoid collisions with other traffic or obstacles.

Instrument Flight Rules

Generally, all flights in IMC must be conducted under IFR, although you can actually fly under IFR at any time, even if the weather is clear – for example, you must obey IFR rules at night. The essential difference between IFR and VFR is that tighter margins are applied for avoiding obstacles and choosing your altitude according to the direction you are flying in (the *Quadrantal Rule* – see *Cruising Altitudes*, below). In Class A, B, C, D, E or F Special Use Advisory airspace, you must also have ATC clearance, and observe any conditions it includes.

Cruising Altitudes

Here, you have to take into account obstacles, aircraft performance and weather. However, what height you fly at is first of all determined by *Quadrantal Rule*, or based on the direction in which you are going, for which, unless told otherwise, you use *magnetic track*.

Quadrantal Rule

This must be observed when outside controlled airspace above the higher of 3000 feet AMSL or the transition altitude, up to FL 240, based on 1013.2 mb. Cruise levels must be at least 1000 feet above the highest obstacle within 5 nm of track.

360-089°	090-179°	180-269°	270-359°
Odds	Odds + 500'	Evens	Evens + 500'

Allowance for Wind Speed

Within 20 nm of ground over 2000 ft amsl, increase safety altitudes by:

Elevation (feet)	0–30 Kts	31–50 Kts	51–70 Kts	+ 70 Kts
2–8000	+ 500'	+1,000'	+1500'	+2000'
+ 8000	+1,000'	+1500'	+2000'	+2500'

This is because the venturi effect over a ridge makes the altimeter misread, as well as causing turbulence and standing waves.

A combination of all this, plus *temperature errors* (see below), can make an altimeter overread *by as much as 3000 feet.*

Temperature Correction

When the surface temperature is well below ISA, correct MSAs by:

Surface Temp (ISA)	Correction
–16°C to –30°C	+ 10%
–31°C to –50°C	+ 20%
–51°C or below	+ 25%

Airspace Structure

Altimeter Setting Regions

When flying through a *Military Aerodrome Traffic Zone* (MATZ), you must set the MATZ QFE.

ATZ

Air Traffic Control Zones are circular, with the centre of the circle based on the longest runway on the relevant aerodrome. If it is 1850 metres or less, the radius of the circle will be 2 nm. Otherwise, it will be 2.5 nm. An ATZ goes up to 2000 feet.

CTA

The lowest level of a CTA (*Control Area*) is at least 700 feet above the surface. Aerodromes underneath a CTA use the same QNH.

CTZ

The lateral limits of a Control Zone extend to at least 5 nm (9.3 km) from the centre of the aerodrome(s) in the directions from which approaches are made.

TCA

A *Terminal Control Area* exists at the confluence of airways or other routes (i.e. where they join) near major aerodromes. Aerodromes under a TCA use the same QNH.

Flight Information Region (FIR)

This is a (generally large) area, within which a flight information and alerting service are provided.

Transition Altitude

This is the altitude at, or below which, any reference your vertical position is based on altitude(which itself is based on QNH). Above the transition altitude, which in UK is generally 3000 feet on QNH, you use *Flight Levels*, based on 1013.2 mb.

When descending to go below transition level, if you are cleared to a Flight Level, you must keep 1013.2 set on your altimeter. If you are cleared to an altitude, and no more FL reports are needed, set the QNH as soon as you start descending and report altitudes.

Near aerodromes, vertical position should be reported in terms of altitude at or below the transition layer, and FLs at or above it.

When passing through the transition layer, report flight levels when going up and altitudes when going down.

Transition Level

This is the lowest available flight level when the altimeter is set to 1013.2 mb, so it would normally be FL 30, including when the QNH is more than standard. However, if the QNH is *less* than standard, the transition level will be *higher*. The difference between the transition altitude and the transition level is the *transition layer*.

ICAO Airspace Classifications

Controlled airspace is classified into *Class A, B, C, D, E, F Special Use Restricted* or *F Special Use Advisory*.

Class A

Separation is provided for IFR aircraft only (VFR is not permitted), from 18,000 feet to FL 600.

Class B

Separation is provided between all aircraft, IFR or VFR, from 12,500 feet (or MEA, whichever is higher) to 17,999 feet. It may contain a control zone and TCA. Clearance is required from VFR aircraft before entering, and position reporting is required, so you need a minimum level of radio/nav equipment. Unless you can get Special VFR, you must leave when conditions demand IFR.

Class C

Separation is between IFR aircraft, with VFR separated from IFR. VFR aircraft require clearance to enter, so you need a 2-way radio. You may get traffic information and conflict resolution. If ATC is not available, it reverts to Class E.

Class D

IFR and VFR, but separation is only between IFR aircraft (traffic information, however, is provided to VFR flights).

Class E

Anything that is still controlled airspace, but not meeting the requirements above, like low level airways, control area extensions, transition areas or control zones without a controlling tower. Separation is as for Class D.

Class F

Where some limitations are imposed. Separation is between IFR aircraft as far as practicable (they receive ATC advisory service) and all flights receive flight information on request.

Class G

Anything not designated as A, B, C, D, E or F, where ATC has no authority, so there's no separation.

Use of Transponder

Unless directed by ATC, when entering UK airspace from a foreign FIR where you don't need a transponder, you must squawk 2000 Mode Alpha, simultaneously with Mode Charlie.

In fact, there are other standard numbers to squawk, when not otherwise instructed, and these are:

- 0000 – malfunction
- 0030 – lost
- 0033 – parachute dropping
- 2000 – from non-SSR area
- 7000 – conspicuity code
- 7004 - aerobatics
- 7007 – open skies

In emergency, squawk:

- 7500 - Hijack

- 7600 – Communications failure

- 7700 - Emergency

Unless authorised, without an SSR transponder, you may not operate in controlled airspace in the UK at or above FL 100.

Radiocommunications

Phonetic Alphabet
To make transmissions clearer if the radios are bad, letters are pronounced in certain ways, as follows:

Letter	Word	Morse
A	Alfa	· —
B	Bravo	— · · ·
C	Charlie	— · — ·
D	Delta	— · ·
E	Echo	·
F	Foxtrot	· · — ·
G	Golf	— — ·
H	Hotel	· · · ·
I	India	· ·
J	Juliet	· — — —
K	Kilo	— · —
L	Lima	· — · ·
M	Mike	— —
N	November	— ·
O	Oscar	— — —
P	Papa	· — — ·
Q	Quebec	— — · —
R	Romeo	· — ·
S	Sierra	· · ·
T	Tango	—
U	Uniform	· · —
V	Victor	· · · —
W	Whiskey	· — —
X	Xray	— · · —
Y	Yankee	— · — —
Z	Zulu	— — · ·

Morse Code
Although the codes (see above) are printed on maps, etc., it's still a good idea to learn them, even if only to keep your job in an airline (many make it a requirement to have at least 6 words a minute). It also stops you peering at your map in the murk and moving your head around too much. Amateur radio clubs are a good source of inexpensive training materials. Starting off at a high speed is best, with the simplest letters. E, for example, is one dot (dit). Just listen to a stream of Morse, picking out that letter only, then add another, such as T, which is a dash (dah), then I (2 dots), M (2 dashes) and so on. In a few days you could be up to 20 words a minute.

Continuous Listening Watch
A listening watch must be kept on appropriate frequencies.

Accidents

A reportable one occurs when:

- anyone is killed or injured from coming into contact with the aircraft (or any bits falling off), including jet blast or rotor downwash

- The aircraft sustains damage or structural failure

- is missing or inaccessible

between the time *any person* boards it with the intention of flight, and *all persons* have disembarked (ICAO definition). This does not include injuries from natural causes, which are self-inflicted or inflicted by other people, or stowaways hiding on places not normally accessible to passengers and crew.

Significant or Substantial Damage in this context essentially means anything that may involve an insurance claim, but officially is damage or failure affecting structure or performance, normally involving major repairs.

Under ICAO, a *fatal injury* is one that involves death within 30 days. A *serious injury* involves:

- more than 48 hours in hospital within 7 days.

- more than simple fractures of fingers, toes and nose.

- lacerations causing nerve or muscle damage or severe haemorrhage.

- injury to any internal organ.

- 2nd or 3rd degree burns or any over 5% of the body.

- exposure to infectious substances or radiation.

The *Accident Investigation Branch* investigates aircraft accidents, and has teams of investigators on 24-hour standby to go worldwide. Its function is not to apportion blame, but to ensure that accidents don't happen again.

An accident must be reported to the AAIB and the local police as soon as possible – the PIC is responsible.

Post Accident Procedures

The pilot or senior survivor, Company or aerodrome authority (in that order, if practical) should take as much as possible of the following action after evacuating passengers to either a sheltered location upwind of the aircraft, or into the liferaft:

- Prevent tampering with the wreckage by ANYBODY except to save life, avoid danger to other persons or prevent damage by fire, for which turn the fuel and battery OFF— disconnect it if there is no risk of a spark, but the AAIB won't like you to touch too much, so remove only emergency equipment, like first aid kits or survival packs, noting where you got them from. Account for all people on board. Attend the injured and cover bodies.

- Activate the distress beacon and maybe use aircraft radio equipment. Prepare pyrotechnics, select, and prepare a helicopter landing site or lay out search and rescue signals.

- If people or communications are close, send for assistance.

- If rescue is likely to be delayed because of distance or failing daylight, prepare suitable shelters, distribute necessary rations of food and water. If necessary, find fresh water.

- Inform the Company (Ops Mgr, Chief Pilot) the quickest way of:

  - Aircraft and Reg No

  - Time and position of accident

  - Details of survivors

  - Nature of occurrence/other details

- Notify Police, Fire, Ambulance, ATC, Gas/Electricity

- Note weather details.

- Make sketches, take photo. Preserve and protect documents and any flight data recorders.

- Refer media to the Company.

Aircraft Accident Reporting

All phone calls and actions taken should be recorded by the person receiving the initial notification – continuous watch should be kept for at least 48 hours or the duration of the process, whichever is longer. Callers should be identified, to ensure it is not a false alarm and to ensure it is indeed a company aircraft. No information should be released without Company authority, mainly for liability reasons.

An Accident Report form should be completed, in addition to complying with the laws and regulations of the country of registration and the country in which the accident or incident occurred. If there is any doubt, the occurrence should be reported as an Accident; it can be reclassified later.

The Company should form an Accident Board, consisting of people with varying qualifications as deemed necessary. This won't be done on the spot, there should be a permanent list somewhere. Only allow 1 photographer and reporter on the scene (let them fight it out amongst themselves).

The accident investigation kit should include a cellphone/satphone, camera, tape recorder, GPS, large-scale map, magnifying glass, compass, tape measure/ruler, plenty of pens and paper (for witness statements and diagrams) first aid kit (to include tweezers), ruler, a packet each of latex and leather gloves, dust

masks, tie tags, surveyor flags and tape, labels, torches, fluid sample bottles, and anything else for the circumstances (duct tape, restricted access signs, etc).

If you get there before the AAIB, take notes, keep detached and don't disturb anything, unless it's going to blow up or catch fire, which would destroy any evidence, including documentation, that needs to be preserved. For photographs, you will need overall scenes, and pictures of gauges, etc. Include anything (such as the ruler in the accident kit) that will indicate scale properly. As for statements, don't put words in witnesses' mouths; just take down what they say.

All documentation relating to the aircraft or pilot should be immediately impounded.

Signals for SAR

You can communicate with SAR aircraft visually by making signals on the ground (the two below are only a selection of the full range available - see the AIP). They should be at least 8 feet high (or as large as possible) with as large a contrast as possible being obtained between the materials used and the background.

```
Need Assistance     V
Need Medical Help    X
```

Air Traffic Services

ATC's mission in life is to *prevent collisions* and *expedite traffic.* They also make every effort to help with rescues and provide information.

The information is disseminated through various offices, including area control centres, terminal control units, control towers, etc.

Here are the services provided:

- *Aerodrome Control Service*, from towers to aircraft and vehicles. The callsign is *Tower* or *Ground*, as appropriate.

- *Area Control Service*, from *Area Control Centres* (ACCs) for flights in control areas. Their callsign is *Control*.

- *Approach Control Service*, for arriving and departing flights. Their callsign is *Approach*. It might also be *Radar*, or *Talkdown* for PAR.

- *Terminal Control Service*, from IFR units (ACCs) or *Terminal Control Units* (TCUs) for IFR and VFR flights in specified control areas.

- *Terminal Radar Service*, an extra from IFR units to VFR aircraft in Class C airspace.

- *Radar Advisory Service*, for information and advisory avoiding action from conflicting traffic. It can be requested at any time, but is usually used in IMC, so you should not accept vectors if they take you there and you are not qualified. This can be time wasting, especially if it's a clear day and you're continually given vectors downwind that take ages to catch up on; although you are not obliged to accept the advice, you must inform the controllers, as you must if you change heading or altitude.

Once advice is refused, you become responsible for traffic

separation, although you are always responsible for obstacle avoidance and obtaining clearances. This can also be expensive, as you become subject to Eurocharges, 100% in UK (but only 25% in France).

- *Radar Information Service*, for informing pilots of the bearing, distance and level of conflicting traffic. Controllers do not offer avoiding action, and updates are only done at the pilot's request if there is a definite hazard. The responsibility for separation is that of the pilot. RIS is normally only available within 30 nm of an Approach radar head.

- *Alerting Service*, for aircraft using ATC, on a flight plan or otherwise, or which are the subject of unlawful interference (officially, the service notifies appropriate organisations about aircraft needing search and rescue, and to assist as required - the *alert phase* is where apprehension exists as to the safety of an aircraft or its occupants). Usually done by a *Rescue Coordination Centre* (RCC).

- *Flight Information Service*, to supply pilots with information about hazardous conditions, especially that which might not have been available on takeoff or have developed since then. The callsign is *Information.*

- *Aerodrome Flight Information Service*, for aircraft near an aerodrome, providing information relevant to the safe and efficient conduct of a flight. It is pre-recorded, and has the prefix *Information,* giving you active runways, pressures, etc.

Clearances, Instructions, etc

You must comply with any *clearance* received and acknowledged. If you don't like it, you should say so at the time, since an acknowledgement without further comment is taken as such. Clearances are valid only in controlled airspace, and there will be some form of the word "clear" in the text to help identify them.

You must also comply with *instructions* in the same way, unless aircraft safety is a factor. An instruction will be readily identifiable, but the word "instruct" may not be included.

If it is not suitable, you may request and, if practicable, obtain an amended one.

Control is based on *known traffic only,* so you are still responsible for safe procedures and good judgement. Any information about flight conditions are meant as assistance or reminders.

Approach Clearances

When you get clearance for an approach, its name will indicate the type of approach if you are required to stick to a particular procedure. If you get visual reference with the ground before completing it, you should carry on with it unless cleared otherwise. If you will be given another runway than that in the approach, the runway number will be given in the clearance as well (in this case, if you have to go around, use the missed approach procedure for the original runway, not the landing one).

Radar Services

Radar allows the best use of airspace by reducing separation between aircraft, and the provision of information, such as traffic and weather. If SSR is available without primary radar, it will not be possible to detect all aircraft.

ATC know which aircraft they are talking to by position reports, identifying turns or transponders. You will be told of any change in your identification status, if any. However, radar identification doesn't stop you being responsible for the disposition of your aircraft, including collision avoidance and obstacle clearance, although ATC accept responsibility for the latter when vectoring IFR flights and certain VFR ones in controlled airspace.

Radar Vectoring

This is used when separation is necessary, for noise abatement, when requested or if an advantage would be gained operationally. You will be told where you are being vectored to, and when it stops, but this can be assumed if you are bound for a final approach or traffic circuit and are given clearance.

Otherwise, it continues until you leave the coverage area, go into controlled airspace or are transferred to a unit that doesn't have radar.

The *minimum radar vectoring altitude* is the lowest that still clears obstacles and is used to make the transition to an approach easier, but it may be lower than the minimum altitudes shown on your chart. If you are cleared to a lower altitude, ATC will be responsible for obstacle clearance until you are in a position to start an approach.

Visual Climb and Descent

If you are being vectored and can see where you are going (that is, you can avoid obstacles yourself and maintain visual reference), you can request permission to climb or descend visually, which may allow you a more direct track. Of course, this means that the responsibility for clearing them is transferred to you, although the proper separation intervals will be maintained.

Pilot Reports

Specialist Aircraft Observations are required whenever severe turbulence or icing is encountered, moderate turbulence, hail or cu-nims are encountered in flight, or anything else that might affect safety.

Intercept Signals

The instructions for these must be carried on all aircraft, but not necessarily as a separate item – they are in the back of the CFS anyway.

Under Article 9 of the Convention on International Civil Aviation, each contracting state reserves the right to stop aircraft from other states flying over parts of its territory. As part of this, aircraft may need to be led away from an area or be required to land at a particular aerodrome.

If an aircraft assumes a position slightly above and ahead of you (normally on the left), rocks its wings, then turns slowly to the left in a level turn, you have officially been intercepted. Your response should be to rock your own wings and follow (the intercepting aircraft will normally be faster than you, so expect it to fly a racetrack pattern and rock its wings each time it passes you). After interception, you should try to inform ATC and try to make contact with the intercepting aircraft on 121.5 or 243 MHz. You may also squawk 7700 with Mode C, unless otherwise instructed.

If the aircraft performs an abrupt break away manoeuvre, such as a climbing turn of 90° or more without interfering with your line of flight, you have been released.

If it lowers its landing gear and descends to a runway (or a helipad), you are expected to land there. You can make an approach to check the area, then proceed to land. Lowering your gear or showing a steady landing light means you acknowledge the instruction. Flashing the landing light means the area is unsuitable, as does overflight with the gear up somewhere between 1000-2000 feet.

At night, the substitute for rocking wings is the flashing of navigation lights at irregular intervals.

International Air Law

The idea behind this is to reduce a phenomenon known as *conflict of laws*, and the resulting confusion that could arise where, say, a claim for damages is brought in a French court in respect of injury to a Canadian whilst travelling on a ticket bought in Holland for a journey from Germany on an Italian plane.

International Air Law has mainly evolved through various International Conventions or Treaties, too numerous to mention here, which form the basis of Public

International Law which in turn can be incorporated into the law of individual states in relation to the Chicago Convention of 1944.

A *Convention* is an agreement that many nations are at liberty to enter into and the word *Treaty* indicates agreements between two (or more) States that bind only themselves. The *Tokyo Convention 1963*, for instance, relates to offences committed on board, but not by, aircraft. Thus, Conventions can cover many subjects, including standards for nav equipment, but they can also establish governing bodies, such as the *International Civil Aviation Organisation* (ICAO).

ICAO is a worldwide body convened by governments while the *International Air Transport Association* (IATA) is an equivalent body established by the airlines. Although IATA is a private organisation, it nevertheless has strong links with ICAO and governments, and is often used by many airlines as an agent for inter-airline cooperation. IATA has many committees, but the most significant is *Traffic*, which negotiates many arrangements between states and airlines. As well as certain freedoms granted by Conventions over the years (such as flying over certain territories, taking tech stops and collecting or discharging passengers), other rights of commercial entry are established by bilateral agreements, which provide for route(s) to be flown, estimate traffic capacity, frequencies of service and establish other precise rules under which operator and crew licensing are accepted by the respective parties to the agreement.

Under the ICAO (Chicago) Convention, the *territory* of a State consists of land areas and adjacent territorial waters under its *sovereignty, suzerainty, protection* or *mandate.* Every State has complete and exclusive sovereignty over the airspace above its territory. Aircraft must comply with the rules of the airspace they occupy, which should be kept as uniform as possible across all states. The responsibility for enforcement lies with the state concerned. Over the high seas, the Convention rules.

The appropriate authorities have the right, without unreasonable delay, to search aircraft belonging to other States when landing or departing, and to inspect documentation. The *State of Registry* is the State in whose Register an aircraft is entered.

Subject to Customs regulations, aircraft on flights to, from or across the territory of another State are admitted temporarily free of duty. Fuel, oil, stores and spares, etc. on board and destined to be leaving again are exempt from duties, inspection fees or similar charges.

Aircraft of other contracting States not on scheduled international air services (i.e. general aviation aircraft) may, subject to the Convention make flights into or non-stop in transit, and to stop for non-traffic purposes (refuel, emergency) without prior permission from the State concerned. However, you may have to follow prescribed routes for safety or security reasons.

Before entering the sovereign airspace of a foreign State with the intention of landing there, the aircraft must be airworthy, with all relevant documents, including the C of A on board.

Crew licences must be issued by the State of registration. They are recognised by other States as long as they exceed ICAO requirements (this also applies to Certificates of Airworthiness).

The Certificate of Registration must be carried at all times.

ICAO Rules of the Air apply to aircraft bearing the nationality and registration marks of a contracting State, wherever they may be, as long as they do not conflict with those of the State with jurisdiction over the territory being flown over. In other words, State rules take precedence over ICAO where they conflict. However, the Authority in the State of registration will most likely require to be informed.

Extracts from the ANO

Exams need you to know about certain parts, so here they are:

Art 3 - Aircraft to be registered

Aircraft flying in or over the UK must be registered somewhere in the Commonwealth, a Contracting State or a country which has a suitable agreement.

Exceptions are kites or captive balloons, and privately operated gliders operating solely in the UK, that is, not engaged on public transport or aerial work, except instruction or flight testing in club aircraft where the instructor and victim, sorry, trainee, are members. In addition, provided the flight takes place only in the UK, an aircraft

without a C of A or permit to fly may fly unregistered for:

- experimental or testing purposes

- qualification for C of A or permit to fly

- demonstrations for sale or to employees

- training or testing of employees

- maintenance flights

The operator and minimum crew must be approved by the CAA and, if not registered, the aircraft must be suitably marked as per Articles 15, 17, 43, 46, 76 and 78. Only essential cargo and persons may be carried. Unless approved, you may not fly over congested areas. See also: *B Conditions in Part A of Schedule 3*.

Art 4 - Registration of aircraft in UK

The CAA is the relevant authority, and maintains the register, but an aircraft will not necessarily be included if it is registered elsewhere and remains so, if an unqualified person holds a share in it, it could be more suitably registered in another part of the Commonwealth, or the public interest is affected.

These people are suitably qualified owners:

- the Crown in right of Her Majesty's Government in UK

- Commonwealth citizens

- nationals of any EEA State

- British protected persons

- bodies incorporated in the Commonwealth with principal places of business in any part

- undertakings formed in EEA States with registered offices, central admin or principal place of business within the EEA

- firms (see Partnership Act 1890) carrying on business in Scotland.

The CAA may, at their discretion, allow an aircraft owned by an unqualified person (living or having a place of business in the UK) to be registered if everything else is OK, provided it is not used for public transport or aerial work, but it may still be registered to a qualified charterer. Otherwise, registration normally becomes void if an unqualified person acquires a legal interest or share, except where an undischarged mortgage is involved, and the other mortgagees disagree. An 'interest in an aircraft' does not include one to which you are entitled only by being in a flying club, and the term 'registered owner' includes, for a deceased person, his legal personal representative, and the successor of a dissolved body corporate.

You must apply in writing to the CAA, and include any relevant information as to ownership. Refer to column 4 of the *General Classification of Aircraft* in Part A of Schedule 2 for its proper description. The register must include:

- certificate number

- nationality mark, and registration mark assigned by the CAA

- constructor's name and its designation

- serial number

- name and address of owners with a legal interest share, or charterer

- an indication that unqualified people are involved

The registered owner should get a certificate of registration, but not necessarily dealers flying for testing, demonstration, delivery or storage (see also Part C of Schedule 2).They should have a place in UK for buying and selling aircraft.

You must inform the CAA in writing of any changes in the original particulars, the destruction of the aircraft, its permanent withdrawal from use, or termination of a demise charter (see above). If you become the owner of a UK registered aircraft, you must inform the CAA in writing within 28 days.

The CAA may amend the register, or cancel a registration within 2 months of being satisfied that there has been a change in ownership, unless it is not in the public interest.

Art 8 - Certificates of airworthiness

Aircraft flying in UK need a C of A from the State of registration, except (when only operating in UK) for kites, or balloons on private flights, privately operated gliders (that is, not on public transport or aerial work, except instruction or flight testing in club aircraft where the instructor and trainee are members), aircraft with a permit to fly, or those working under A and B Conditions in Part A of Schedule 3, where, generally speaking, the operator and minimum crew must be approved by the CAA and, if not registered, the aircraft must be suitably marked, complying with Articles 15, 17, 43, 46, 76 and 78. Only cargo and persons essential to the flight may be carried. Unless approved, you may not fly over congested areas.

With UK registered aircraft, the issue of a C of A depends on the design, construction, workmanship and materials (including engines) and any necessary equipment for airworthiness, together with the results of flying trials, or other tests, which may be dispensed with if a prototype has had them done already.

The C of A specifies the appropriate categories under Part B of Schedule 3, and is issued under the assumption that the aircraft is flown for the purposes indicated. It also specifies the performance group for Article 36(1). It remains in force for the period specified, or until the aircraft (or necessary equipment) is overhauled, repaired or modified, or equipment removed or replaced in an unapproved fashion, or the completion of mandatory servicing or modifications.

A foreign C of A may be validated at any time.

Art 11 - Technical Log

Required by every UK registered aircraft with a transport or aerial work category C of A, unless it is below 2730 kg MAUW and not operated by an AOC holder, where an approved alternative may be used. It must be filled in at the end of every flight, by the commander,

specifically noting the takeoff and landing times, any known defects that affect airworthiness or safe operation (or *nil defects*) and anything else required by the CAA from time to time. Entries must be signed and dated. Those relating to defect rectification must be readily identifiable with the defect concerned. For a series of consecutive flights on the same day at the same aerodrome with the same commander (except spraying), the details may be filled in at the end of the last flight unless a defect must be reported.

Tech logs must be carried in the aircraft (see Article 76), with copies of the entries above kept on the ground, except for helicopters and aeroplanes weighing less than 2730 kg, where they may be carried on board in an approved container, if it is not reasonably practicable to do otherwise. They must be preserved for 2 years after the aircraft has been destroyed or permanently withdrawn from use, unless Article 80 disagrees, or the CAA varies it.

Art 12 - Inspection, overhaul, repair, replacement and modification

For any UK registered aircraft requiring a C of A, unless it comes under JAR-145.

Except for special category aircraft below 2730 kg, or private or special category aircraft where the owner/pilot does the work (in which case suitable records must be kept and equipment used), a Certificate of Release to Service must be valid for the aircraft itself, and any necessary equipment or radios. The certificate must certify that any inspection, overhaul, repair, replacement, modification or maintenance has been done in an approved manner with approved materials, and include particulars. It may also only be issued by the holder of an AME licence under JAR or CAA (or validated foreign). When abroad, you can use a foreign engineer if the MAUW is less than 2730 kg.

A UK or JAR ATPL holder, or UK Flight Navigator, may adjust direct reading magnetic compasses (this comes within the definition of 'repair').

If work is done where the standard is low, or a certificate cannot be issued, you can fly to the nearest place you can safely get a certificate, always ensuring you write and tell the CAA about the whole affair within 10 days.

Aircraft log books must be kept for 2 years after the aircraft has been destroyed or permanently withdrawn from use (but see also Article 80).

Art 14 - Equipment of aircraft

Aircraft must be equipped under the law of the country in which they registered, and must be able to display lights and markings, and make signals, under this and any subsequent orders.

UK registered aircraft need equipment in Schedule 4, which must be approved, except for certain items, such as timepieces – see paragraph 3. The CAA may vary this at any time in particular cases. Public transport aircraft must have, for each passenger, a flight briefing card, with the usual information on it.

Equipment must not be a danger in itself, and emergency equipment must be clearly marked.

This does not apply to radio apparatus, except that in Schedule 4 (see below).

Art 15 - Radio equipment of aircraft

Aircraft need radios and radio navigation equipment under the law of the country of registration, or the State of the operator. Refer to Schedule 5, which may be varied at any time by the CAA.

Equipment and installation methods must be approved for UK registered aircraft, except unregistered gliders under Art 3(2).

Modifications must not be made without CAA approval.

Art 17 - Aircraft, engine and propeller log books

UK registered aircraft must have log books for the aircraft itself, each engine and variable pitch propeller. See Schedule 6 for what must be in them. If the aircraft is below 2730 kg, the log books must be approved by the CAA.

Entries must generally be made as soon as practicable after the occurrence, but no later than 7 days after the expiration of the current certificate of maintenance review. However, those required by sub-paragraphs 2(d)(ii) or 3(d)(ii) of Schedule 6 must be made at the time (e.g. inspections, overhaul, etc).

Documents mentioned in log books are regarded as part of them.

Log books must be kept until 2 years after the aircraft, engine or variable pitch propeller has been destroyed or has been permanently withdrawn from use, but see also Article 80.

Art 18 - Aircraft weight schedule

Every flying machine and glider with a C of A must be weighed, and its C of G determined. A weight schedule must be prepared, showing either the basic weight (e.g. empty with unusable fuel and oil, and equipment in the weight schedule) and the C of G for the basic weight, or approved alternatives.

Weight schedules must be kept for 6 months after the next weighing, but see Article 80.

Art 20 - Composition of crew

Aircraft must carry flight crews as required by the country of registration. Crews on UK aircraft must be adequate for safety and be at least as specified in the C of A.

A UK registered public transport machine over 5700 kg must have at least two pilots, as must turbo jets, pressurised turbo props, and multi-engined turbo props able to carry more than nine passengers below 5700 kg under IFR. Under the same conditions, multi-engined turbo props able to carry less than 10 passengers without pressurization or multi-engined pistons may use an autopilot instead (those under a police AOC may fly single pilot, as may the multi-engined turbo props and pistons mentioned above)

A UK registered public transport helicopter below 5,700 kg able to carry 9 or less passengers under IFR or at night must carry two pilots, unless it has an autopilot with at least altitude hold and heading

mode, or is working under a police AOC.

A UK registered public transport aircraft must either carry a flight navigator or suitable navigation equipment if it is planned to be more than 500 nm from the point of take-off and passing over part of any area in Schedule 7. The flight navigator is in addition to anyone carried under this article for other duties.

A UK registered aircraft required by Art 15 to have radio communications apparatus must carry a flight radiotelephony operator.

The CAA may specify additional crew members at any time.

A UK registered public transport aircraft carrying more than 20 passengers, or capable of carrying 35 but carrying at least one, must carry cabin crews. The ratio is one to every 50 (or fraction of 50) seats, unless varied by the CAA.

Art 21 - Requirement for licence

Flight crews need appropriate licences. Having said that, within the UK, the Channel Islands, and the Isle of Man, you can operate radios without a licence as the pilot of a UK registered glider not flying for public transport or aerial work and not talking to ATC, or being trained in a UK registered aircraft for flight crew duties, when authorised by a licence holder. The messages must only be for instruction, or the safety or navigation of the aircraft, on frequencies exceeding 60 MHz assigned by the CAA, which are automatically maintained, and if the transmitter has external switches.

For other licences, where an aircraft is being towed, no-one else must be on board unless they are authorised or are a trainee. Ex-military pilots (within the last 6 months) or flight engineers/navigators upgrading to pilot or adding types may also fly without licences under supervision. Military pilots may fly without licences in the course of their duties.

You can also be a PIC for renewal of a pilot's licence or inclusion or variation of ratings if you are at least 16 with a valid medical certificate (and comply with its conditions), under the supervision of an instructor and nobody else is carried. It must also be a non-public transport or aerial work flight, although your instructor cab get paid. On single-pilot aircraft, dual controls must be fitted for the instructor.

You can act as PIC of a helicopter or gyroplane at night if you do not have an IR and have not carried out at least 5 takeoffs and landings at night in the last 13 months, under an instructor, with nobody else on board, and not engaged on public transport or aerial work, except paying the instructor. The same conditions apply to balloons if you have not done 5 flights over 5 minutes.

A JAR licence is considered valid for this article. A non-JAR foreign licence does not allow you to be paid for public transport or aerial work, fly IFR, or give instruction.

You don't need a licence to fly a glider unless you are operating the radio or you are engaged on public transport or aerial work, other than aerial work consisting of instructing or testing in a club aircraft and the people concerned are members.

You need permission from a Contracting State if you fly in or over it in a UK registered aircraft and your (UK) licence is endorsed as not meeting the full specification. You also need permission from the CAA if your foreign licence is not up to scratch and you fly in or over UK.

Refer also to Article 26 (3) concerning medical fitness and pregnancy.

Art 41 - Pilots to remain at controls

During flight, one pilot must remain at the controls of UK registered flying machines or gliders. If two are required, both must be there during takeoff or landing. With two or more on public transport, the commander must be there. Each pilot must wear a safety belt with or without one diagonal shoulder strap, or a safety harness, except that, during take-off and landing a safety harness shall be worn if it is required by article 14 of and Schedule 4.

If you ever have to leave the controls of a helicopter with the engine running, *do not* switch the hydraulics off, but use the control locks only, in case the controls motor by themselves.

Art 45 - Duties of commander

Commanders of UK registered public transport aircraft (except police aircraft) must take all reasonable steps to do the following:

In a landplane intending to fly more than 30 minutes away from land, or gliding distance if cabin crews are carried (and there is a possibility of having to land on water during takeoff or landing, including alternates), all passengers must be shown how to use the lifejackets. If the landing on water arises from going to an alternate, the demonstration may be delayed until after the decision has been made to go there. In a seaplane, ensure that all passengers are shown how to use the above equipment before takeoff.

Before the aircraft takes off, and before it lands, ensure that the crew are properly secured in their seats, together with anyone carried under article 20(7) (cabin crews), who must be situated so they can help passengers.

For takeoff and landing, and turbulence, ensure that all passengers over 2 years old are wearing seat belts (including diagonal shoulder straps) or safety harnesses and that all passengers under 2 years are using a child restraint device.

Ensure that baggage in the passenger compartment is secured and, for aircraft capable of seating more than 30 passengers, it is either stowed in approved spaces or elsewhere with written permission.

For unpressurised aircraft with a C of A issued on or after 1st Jan 1989, before FL 100, show passengers how to use the oxygen. Above FL 120 all passengers and cabin crews should be using it, with all flight crew using it above FL 100. Where the C of A was issued before then, you have the choice of the above or doing the demonstration before FL 130, with cabin crews and passengers using it above that.

Art 46 - Operation of radio in aircraft
Radios must be operated under the terms of any licence issued by the country of registration, or the State of the operator, and by licensed people at that, as instructed by ATC or relative to the airspace. They must not cause interference and may only be used according to general international aeronautical practice.

Where radios are required, a continuous listening watch must be kept by the a member of the flight crew. You may change frequencies when so directed. If there is no objection, you can use a device to listen for you.

Hand-held microphones may not be used in UK registered public transport aircraft in controlled airspace below FL 150, or during takeoff or landing.

Art 53 - Flight recording systems
See paragraph 4(4), (5), (6) or (7) of Schedule 4 to see if CVR or FDR is required. If so, they must be in use from the start of takeoff to the end of landing, or from rotor start to stop in a helicopter. The last 25 hours of an FDR must be preserved, together with a record of at least one representative flight in the last 12 months, to include a take-off, climb, cruise, descent, approach to landing and landing, plus a means of identifying the record. For a helicopter, you need the last 8 hours of an FDR. If combined with a CVR, then you can choose either the last 8 hours, or the last 5 hours against the duration of the last flight, whichever is greater, plus the period immediately preceding the last 5 hours or duration of the last flight

(whichever is the greater) or whatever the CAA may dictate.

See also Article 80.

Art 54 - Towing of gliders

The C of A of a towing aircraft must allow this. The combined length of towing aircraft, tow rope and glider must not be more than 150 metres.

The tow rope must be in good condition and strong enough, and the performance of the towing aircraft good enough to take off safely, reach and maintain a safe height for separation and land safely afterwards at the intended destination.

Signals must be agreed and communication established with ground crew, together with emergency signals between the commander of the towing aircraft and that of the glider, that the tow should immediately be released by the glider, and that it cannot be released.

The glider must be attached to the towing aircraft before take off.

Art 55 - Towing, picking up and raising of persons and articles

Except for radio aerials, experimental instruments, signals, apparatus or other required or permitted article, in emergency or saving life, or when flying under B Conditions in Part A, the C of A of the aircraft concerned must allow this (but see article 54 for gliders).

Launching or picking up of tow ropes, banners or similar articles must be done at an aerodrome. Except for gliders, towing may not be done at night or when visibility is less than one nautical mile. The combined length of towing aircraft, tow rope and whatever is being towed must not be more than 150 metres.

A helicopter may not fly over a congested area of a city, town or settlement with a slung load. Only essential people may be carried.

Art 56 - Dropping of articles and animals

Articles and animals (whether or not by parachute) shall not be dropped (or projected or lowered) from an aircraft in flight so as to endanger persons or property, or at any time to the surface, unless under an aerial application certificate (see article 58), for saving life, jettisoning in emergency, as ballast in the form of fine sand or water, navigation purposes, tow ropes or banners at aerodromes, public health, pollution or weather purposes (with permission) or wind drift indicators for parachutists (also with permission).

This does not include lowering an article or animal from a helicopter to the surface, assuming its C of A allows it.

Art 57 - Dropping of persons

Except in emergency, or when saving life, you are not allowed to drop, be dropped or permitted to drop to the surface or jump from an aircraft flying over the UK except under a police AOC or written permission from the CAA. This includes projecting and lowering. In any case, it cannot be done while causing danger to persons or property.

It must be allowed by the C of A, or be done under a police AOC.

If you have permission, you must have a parachuting manual, as amended, which serves the same purpose as an ops manual.

Art 59 - Weapons and munitions
These may not be carried without permission from the CAA, and the commander informed in writing (unless under a police AOC) by the operator before flight of its type, weight, quantity and location.

Sporting weapons or munitions of war (including parts) in any case may not be carried where passengers have access, except for police aircraft. They may only be carried on board as checked baggage or consigned as cargo, or in a place inaccessible to passengers, and must be unloaded. The operator must be told about it beforehand and must consent.

This does not include weapons taken on board for aircraft safety.

Art 60 - Dangerous goods
This just allows the Secretary of State to make regulations, with the provisions. They are on top of anything in Article 59 (munitions of war, sporting weapons).

Art 61 - Carriage of persons
Except for police helicopters (with special procedures in place), people may not be carried in flight where they are not supposed to be, particularly on the wings or undercarriage. Neither may they be in or on anything towed, except for pilots in gliders and flying machines. However, temporary access may be allowed for the safety of the aircraft or any person, animal or goods in it, and any part where cargo or stores are carried, if designed for it.

Art 63 - Endangering safety of an aircraft
You may not recklessly or negligently act so as to endanger an aircraft, or any person in it.

Art 64 - Endangering safety of any person or property
You may not recklessly or negligently cause an aircraft to endanger people or property.

Art 65 - Drunkenness in aircraft
You may not enter any aircraft when drunk, or be drunk in any aircraft. Flight crew members may not be under the influence of drink or a drug to such an extent as to impair their ability to act as such.

Art 66 - Smoking in aircraft
Notices, visible from each passenger seat, indicating when smoking is prohibited must be exhibited in every UK registered aircraft.

No smoking may take place in any compartment of a UK registered aircraft when it is prohibited there by a notice exhibited by or on behalf of the commander.

Art 67 - Authority of commander and members of the crew of an aircraft
Every person in an aircraft must obey all lawful commands given by the commander for securing its safety and that of persons or property in it, or the safety, efficiency or regularity of air navigation.

Art 70 - Flying Displays

Except for the military, or aircraft races, organisers (i.e. flying display directors) must get CAA permission (in writing) for the display. Applicants must be fit and competent, with regard to previous conduct and experience, organisation, staffing and other arrangements, to safely organise it.

Commanders of aircraft intending to participate must take all reasonable steps to check that such permission has been obtained, the flight can comply with any conditions given with it, and that an appropriate pilot display authorisation has been granted. Applicants for authorisations must be suitable as above to safely fly in the display. Usually, a JAA authorisation is valid, but the CAA may change this.

There may be special conditions for military aircraft in a civilian display.

Art 76 - Documents to be carried

Aircraft must fly with documents prescribed under the law of the country of registration. However, UK registered aircraft must carry those under Schedule 11, except when taking off and landing at the same aerodrome within the UK, where they may be left there.

Art 78 - Production of documents

You must, within a reasonable time after a request from an authorised person, produce the C of A and C of R, flight crew licences, and any documents mentioned in article 76 (i.e. in Schedule 11). This request must obviously be made personally, because you will need to check their credentials, and because a request to send your licence somewhere is

actually one to "surrender", which is not the same thing according to the 1982 Act. The Act itself specifically mentions "custody", "production" and "surrender" as three separate things, so the intention clearly is to regard them as such. A "reasonable time" is as long as it takes to reach inside your navbag in an immediate post- or pre-flight situation, or within five days of the original request, like for driving licences.

As the Civil Aviation Act only allows provision by the ANO for access to *aerodromes* and *places where aircraft have landed* for the inspection of documents, it's arguable that requests for production are invalid if done at your home, for example. A constable, of course, can go anywhere within the limits settled by the Act and the ANO.

An operator of a UK registered aircraft must, within a reasonable time, etc. etc., produce:

- Radio licence, C of A and C of R (Schedule 11 A, B & G)

- the aircraft, engine and variable pitch propeller log books

- the weight schedule

- for public transport or aerial work aircraft, loadsheet and relevant parts of ops manual (Schedule 11 D and H)

- for aircraft with a transport or aerial work category C of A, certificate of maintenance review and tech log (Schedule 11 E and F)

- records of flight times, duty and rest periods, and any associated with them

- any operations manuals

- the record made by any FDR

Holders must produce any licences within a reasonable time after a request from an authorised person, including validations and medical certificates.

A personal flying log book must be produceable within a reasonable time after a request from an authorised person for 2 years after the date of last entry.

Art 83 - Offences in relation to documents and records

You may not, with intent to deceive, use a forged, or otherwise altered document (or copy), or one to which you are not entitled. Neither may you allow it to be used by anyone else, or tell lies in order to get one.

You may not intentionally damage, alter or render illegible any log book or other record required to be maintained, or knowingly make false entries in or material omissions from them, or destroy them while they are meant to be preserved. The same goes for loadsheets.

Written entries must be in ink or indelible pencil. All statements must be correct.

No unauthorised certificates under JAR-145 may be issued.

Art 84 - Rules of the Air

This article gives the Secretary of State permission to make The Rules of the Air, concerning how aircraft may move or fly, including giving way to military aircraft, lights or signals to be shown or made, lighting

and marking of aerodromes and anything else concerning safety.

This paragraph also makes it an offence not to obey the Rules, unless you are trying to avoid immediate danger, or are complying with the laws of any country you may be in, or doing anything with a military aircraft in the normal course of your duties (e.g. under JSP 318 or Flying Orders to Contractors issued by the Secretary of State).

If you disobey the Rules to avoid immediate danger, you must report it within 10 days to the competent authority of the country where the incident happened or, if over the high seas, to the CAA. However, the Rules don't absolve you from any neglect in using lights or signals or any precautions required by ordinary aviation practice, or by any special circumstances.

Art 85 - Power to prohibit flying

The Secretary of State may restrict or prohibit flying for the intended gathering or movement of a large number of persons, the holding of an aircraft race, contest or flying display, national defence or any other reason affecting the public interest. Regulations may also be made prohibiting, restricting or imposing conditions on flight by any aircraft over the UK or near an offshore installation, and any UK registered aircraft, in any other airspace for which Her Majesty's Government provides aerial navigation services.

As soon as you become aware that you are in any restricted airspace established for national defence or in the public interest, you must leave by the shortest possible route, and not descend, unless otherwise instructed. You must obey any instructions given by ATC or whoever is responsible for safety inside the airspace, including Danger Areas.

Prohibited Airspace

This may only be created by the ANO itself; the 1982 Act permits aircraft to be stopped from flying over such areas as may be specified therein. It also allows the ANO to provide for exemptions, so, unless specified in the ANO, or exempt under the terms given in it, prohibited airspace does not exist. Mere "notification" of its existence may not be enough.

Art 86 - Balloons, kites, airships, gliders, parascending parachutes

This only applies to aircraft in the UK.

Without permission from the CAA, a captive or tethered balloon may not fly within 60 metres of any vessel, vehicle or structure, except with the permission of any person in charge of them, with the top of the balloon not more than 60 metres above ground level. A captive balloon may not fly within an aerodrome traffic zone during its notified operating hours. Neither can a kite above 30 metres from ground level and, in any case, above 60 metres. A glider or parascending parachute may not be launched by winch and cable or by ground tow to more than 60 metres above ground level, with the latter not being allowed to fly within an aerodrome traffic zone during its notified operating hours.

An uncontrollable balloon needs CAA permission to fly in airspace notified for this paragraph. A controllable balloon in free controlled flight can only fly in such airspace, or within an aerodrome traffic zone during its notified operating hours during the day in VMC. In tethered flight, it needs permission from ATC.

Captive balloons must be securely moored and not left unattended unless it has a means of automatic deflation if it breaks free.

An airship over 3000 cubic metres may only be moored at a notified aerodrome, except with written permission from the CAA. An airship under 3000 cubic metres, unless moored on a notified aerodrome, may not be moored within 2 km of a congested area, or in an aerodrome traffic zone, except with written permission from the CAA. An airship in the open must be securely moored and not left unattended.

You must give the CAA 28 days' notice in writing to release groups of small balloons over 1000 in size simultaneously (i.e. within 15 minutes) at a single site (inside 1 square km) within an aerodrome traffic zone during its notified operating hours. For between 2,000-10,000, or over that, you need written permission.

Art 103 - Licensing of aerodromes

A licence may be granted on the basis that the applicant is competent, the aerodrome is safe and the aerodrome manual is adequate. A *licence for public use* means it is available to all persons on equal terms and conditions, in which case the times during which it is open for public transport or flying instruction must be notified. The licence does not become invalid if article 101(2) is contravened.

An aerodrome manual must be submitted with every application, and relevant parts must be made available to all operating staff. Amendments must be furnished to the CAA before or immediately after they come into effect.

Art 112 - Aviation fuel at aerodromes

Installations must not render fuel unfit for use in aircraft. They must be appropriately marked and sampled, with written records kept for at least 12 months. This does not apply to fuel removed from one aircraft and meant for use in another from the same operator.

Art 122 - Penalties

The operator, commander and charterer (for article 113) may be liable for contraventions of the ANO, including public transport, unless it can be proved that they were done without their consent or connivance, and that all due diligence was performed to prevent them, or they could not have been reasonably avoided.

Usually fines on summary conviction will not exceed Level 3 on the standard scale. However, Part A of Schedule 12 rates Level 4, and Part B gets you the maximum on summary conviction and a fine or two years' porridge (or both) on conviction.

Art 129 – Interpretation

See *Glossary*.

Some Questions

1. An aircraft overtaking another in flight must pass to which side?

2. An aircraft's vertical position with the altimeter set to 1013.2 mb (29.92") is reported as what?

3. The Transition Altitude is that at or below which vertical position is controlled by reference to what?

4. What is the minimum radio equipment for Class D airspace?

5. Who is responsible for the safe conduct of a VFR flight?

6. What is an FIR?

7. What is a Control Zone?

8. Is an airway a Control Zone or a Control Area?

9. How is separation provided?

10. What is the minimum time for filing a flight plan?

11. What is a visual contact approach?

12. On what occasions would you consider diversion?

13. What qualification is required for IFR flight in controlled airspace?

14. In IMC at FL 90, what should your magnetic track be?

Some Answers

1. The Right.

2. A Flight Level.

3. Altitude.

4. VHF Comms.

5. The pilot.

6. Flight Information Region, a large area (there are two: London & Scottish) extending up to but not including FL 245.

7. Notified airspace starting at ground level in which ATC service is given to IFR flights, so when VFR you can get away with ATC control, unless it is notified under Rule 21, where you need permission to enter in the first place.

8. A Control Area, since it does not start at the ground.

9. Track and Geographical.

10. 30 mins.

11. An instrument approach with all or part of it completed by visual reference to terrain.

12. Weather below minima, runway obstructed, failure of ground services and unacceptable delays.

13. An Instrument Rating.

14. Between 000-089 degrees.

Human Factors

Aircraft are getting more reliable so, in theory at least, accidents should happen less often. Unfortunately, this is not the case, so we need to look somewhere else for the causes. Believe it or not, accidents are very carefully planned – it's just that the results are very different from those expected! An accident is actually the end product of a chain of events, so if you can recognise the sequence it should be possible to nip any problems in the bud.

A common saying is that "the well oiled nut behind the wheel is the most dangerous part of any car". Not necessarily true for aviation, perhaps, but, in looking for causes other than the hardware when it comes to accidents, it's hard not to focus on the pilot (or human factor) as the weak link in the chain— around 75% of accidents can be attributed to this, although it's also true to say that the situations some aircraft are put into make them liable to misfortunes as a matter of course, particularly helicopters – if you continually land on slippery logs in clearings, something untoward is bound to happen sometime!

The trend towards human factors in relation to accidents was discovered through the 80s and 90s, when a series of accidents that occurred in the USA were analysed in depth. It was found that crew interaction was a major factor since, nearly 75% of the time, it was the first time they had flown together, and nearly half were on the first leg, in situations where there was pressure from the schedule (over 50%) and late on in the duty cycle, so fatigue was significant. The Captain was also flying 80% of the time. The problem is, that it's not much different now – 70% of accidents in the USA in 2000 were pilot related, based on mistakes that could easily be avoided with a little forethought. Now, the figure worldwide is around 80%. If air traffic continues to grow at the present rate, we will be losing 1 airliner per week by 2010.

Since the problem of crew co-operation needed to be addressed, management principles used in other industries (i.e. Quality Assurance and Risk Management) were distilled into what is mostly called *Crew Resource Management*, triggered by three accidents, one of which was at Dryden, also instrumental in new Canadian icing laws being passed (in Canada, a condensed version concerned mainly with *Pilot Decision Making* is needed to fly in visibility down to half a mile. This might not actually be due to weather, but smoke, as you might find in a forest fire – in fact, you can be nearly IMC on a hazy day in some industrial areas. However, this "licence" for bad weather flying does not mean you have to do it – it's not the equivalent of the amber traffic light meaning "go faster"! You still have to be aware of the implications of what you are doing).

In other words, if you have a pilot behaving under par in an aircraft where it shouldn't be, you're just asking for trouble, and this applies to large aircraft just as much as it does to small ones. An accident-prone person, officially, is somebody to whom things happen at a higher rate than could be statistically expected by chance alone. Taking calculated risks is completely different from taking chances. Know your capabilities, and your limits.

Bad weather visibility is associated with low ceilings, and familiarity with the area is a real help, so local flying is better than a low-level navex, at least without a GPS. This, at least, will save you changing your focus from the outside to the map inside your cockpit, which is not where it should be in such circumstances. However, GPSs produce their own problems – because they help you so much when the weather is bad, they tempt you to stretch the envelope, which is dangerous in itself.

Most weather-based accidents involve inadvertent entry into IMC by people who have only had the basic instrument instruction required for the commercial licence. Next in line is icing.

Previously, you might have been introduced to the concept of *Airmanship*, which involved many things, such as looking out for fellow pilots, doing a professional job, not flying directly over aircraft, etc. – something that could be called being the "gentleman aviator".

These days, there are new concepts to consider, such as delegation, communication, monitoring and prioritisation, although they will have varying degrees of importance in a single-pilot environment. In fact, the term "pilot error" is probably only accurate about a third of the time; all it really does is indicate where a breakdown occurred. There may have been just too much input for one person to cope with, which is not necessarily error, because no identifiable mistakes were made. Perhaps there needs to be a new phrase, occupying the same position that "not proven" does in the Scottish Legal System, which lies between Guilty and Not Guilty.

The aim of this sort of training is to increase flight safety by showing you how to make the best use of resources available to you, which include your own body (physical and psychological factors), information, equipment and other people, whether in flight or on the ground— P2s are trained for emergencies, for example, so they can be used instead of automatically taking over yourself when something happens – like a human autopilot, in fact. Also remember that the behaviour of people in a company is very much a reflection of the management, in our case the commander, so there is an obligation for whoever's in charge to foster a positive working environment, which, essentially, means not being miserable. Like it or not, you are part of a team, even if you are the only one in the cockpit, and you have to fit into an established system. All of this is geared to help you with making decisions, of which more later.

The aim of a PDM course, in particular, is to help pilots make better decisions by introducing them to the concepts, principles and practices of good decision-making, with the intention of reducing the accident rate even further. That is to say, we know all about the hardware, now it's time to take a look at ourselves.

It has been noticed that pilots who receive decision-making training outperform others in flight tests and make 10-15% fewer bad decisions, and the results improve with the comprehensiveness of the training.

The courses are supposed to be discussion-based, which means that you are expected to participate, with the intention that your experiences will be spread around to other crews. This is because it's quite possible never to see people from one year to the next in a large organisation, and helicopter pilots in particular have no flying clubs, or at least opportunities to "hangar fly" as the Americans say, so experience is not

being passed on. In fact, even if you operate out in the bush, you might see some of your colleagues at the training sessions at the start of the season, and not see them till the end.

One accident which illustrates the need for CRM training was a Lockheed 1011 that flew into the Florida Everglades. A problem involving the nosewheel occupied the attention of all three members of the crew so much that they lost the big picture, and the aircraft ended up in the swamp. It was concluded that the commander should have ensured that someone was monitoring the situation, and should have delegated tasks accordingly. But was a "mistake" actually made? Nobody pushed the wrong switches or carried out the "wrong" actions – it was maybe just a wrong decision.

A contribution to the Kegworth accident in UK, where the plane ended up on the motorway, was the inability of the cabin crews to feel they were able to talk to the flight deck if they saw a problem, which puts the problem fairly and squarely at the door of the Company, or at least the management. Also, a reading of the accident report on the Air Florida flight that hit a bridge and ended up in the Potomac would be instructive—the FO was clearly sure that something was wrong (icing) but didn't like to say so.

In short, *Crew Resource Management* (CRM) is the effective utilisation of all available resources (e.g. crew members, aeroplane systems and supporting facilities) to achieve safe and efficient operation - the idea is to enhance your communication and management skills in order to achieve this. In other words, the emphasis is placed on the non-technical aspects of flight crew performance (the so-called *softer skills*) which are also needed to do your job properly. As we said before, you could loosely call it airmanship, but I prefer to use the term *Captaincy*, as flying is a lot more complex now than when the original term was more appropriate.

The elusive quality of Captaincy is probably best illustrated with an example, such as the subject of the Critical Point. If you can think back to your pilot's exams, you will recall that it is a position where it takes as much time to go to your destination as it does to return to where you came from, so you can deal with emergencies in the quickest time.

In a typical pilot's exam, you will be given the departure and destination points, the wind velocity and other relevant information and be asked to calculate the CP along with the PNR (*Point of No Return*), which is alright as far as it goes, but tells you nothing about your qualities as a Captain, however much it may demonstrate your technical abilities as a pilot.

Now take the same question, but introduce the scenario of a flight across the Atlantic, during which you are tapped on the shoulder by a hostess who tells you that a passenger has got appendicitis. First of all, you have to know that you need the CP, which is given to you already in the previous question. Then you find out that you are only 5 minutes away – technically, you should turn back, but is that really such a good decision? (Actually, it might be, since it will take a few minutes to turn around). Commercially, it would be

disastrous, and here you find the difference between being a pilot and a Captain, or the men and the boys, and why CRM training is becoming so important.

A Captain is therefore supposed to exhibit qualities of loyalty to those above and below, courage, initiative and integrity, which are all part of the right personality – people have to *trust* you. This, unfortunately, means being patient and cheerful under the most trying of circumstances, and even changing your own personality to provide harmony within the crew, since it's the objective of the whole crew to get the passengers to their destination safely.

With regard to outside agencies, as single crew, there is only you in your cockpit, but you still have to talk to passengers and others in your organisation, and we all work in the Air Transport Industry. It just happens that your company is paying your wages at the moment—in this context, the word "crew" includes anybody else who can help you deliver the end product, which is:

.. Safe Arrival!

Very few people travel just for the sake of it, unless you own a Pitts. Everything else is subordinate to this, including pride and increasing your qualifications and experience. Remember that the general public are paying your wages.

ATC are there to help if you've got a problem. They will check your landing gear is down, file a flight plan and check out the weather, although a lot of this could be done by the Company if they've got a radio frequency.

The best way out of trouble is not to get into it, which is easier said than done with an intimidating passenger or management. You, the pilot, are the decision-maker – in fact, under the Chicago Convention (Annex 2, Chapter 2), your word is law until overturned within 3 months by a person with a lawful interest. However, the other side of the coin is that you are liable for what goes on – in fact, in aviation, the buck very definitely stops at the bottom.

To do this properly, we need to look at Man, the machine and the environment - you know about the machine – we will therefore concentrate on the other two.

Decisions, Decisions

Your licence means the authorities consider you to have enough training to make decisions – however, it cannot, and does not, cater for every situation.

Aviation is noticeable for its almost constant decision making. As you fly along, particularly in a helicopter, you're probably updating your next engine-off landing point every five seconds or so. Or maybe you're keeping an eye on your fuel and continually calculating your endurance in view of unexpected weather. It all adds to the many tasks you're meant to keep up to date with, because the situation is always changing. In fact, a decision not to make a decision (i.e. wait a while for developments) is also a decision, always being aware that we don't

want indecision. To drive a car 1 mile, you must process 12,000 pieces of information – that's 200 per second at 60 mph! It has to be worse with flying, and possibly over our limits – we can begin to see that our capability of processing information is actually quite marginal.

A decision is actually the end result of a chain of events involving judgment. The process involves not only our eyes and ears which gather data, but our attention, which should not be preoccupied all the time. The human body is not multi-tasking, and to keep track of what's going on it's necessary to split your attention for a short period between everything; typically a split second at a time – although decision making is a process involving several steps, things happen simultaneously, so it's important not to get fixated on one thing at the expense of another, which is typically what happens when flying in bad weather (remember the 1011). Gather all the information you can in the time available or, better still, get in the habit of updating information you're likely to need in an emergency as the flight progresses.

In flight, you take on the role of an information processor – in this, you has a unique talent, in that a decision can be made without all the relevant information. If you were to ask a computer to choose between a clock that was gaining five minutes a day, and one that had stopped completely, it would probably choose the one that had stopped, because it was accurate twice a day, as opposed to once every 60 days or so. The point is that machines cannot discriminate, and they need *all* irrelevant information, which is good if you just want them to report facts, as with instruments, but not if you want them to make decisions.

In fact, the steps are to:

- Gather relevant information – using your senses (which may be wrong)

- Process it – keeping situational awareness (i.e. keep the big picture going and don't get fixated)

- Make a decision – but other factors may affect its quality

- Act on that decision – although other factors may affect your ability to implement it

The above steps are not rigid, but may be merged or even repeated during a situation. It can be made quicker if some experience has already been gained, hence the value of training. You gather information through the senses, but these don't always tell the truth, a subject we will look at later (of course, the information itself may be wrong).

Information Processing

Physical stimuli, such as sound and sight, are received and interpreted by the brain. *Perception* at this point means converting that information into something meaningful, or realising that it's relevant to what you're doing. What comes out depends on past experience of those events, your expectations, and whether you're able to cope with the information at that time. Good examples are radio transmissions from ATC, which you can understand, even if you can't hear

them properly, because you expect certain items to be included, and you know from experience that they're bad anyway. The danger, of course, is that you may hear what you want to hear and not what is actually sent! (see *Communication*, below). We can only do this one item at a time, however.

Each decision you make eliminates the choice of another so, once you make a poor one, a chain of them usually follows. In fact, a decision-making chain can often be traced back up to and over fifty years, depending on whether the original cause was a design flaw. Another factor is the data itself; if it's incomplete, or altered through some emotional process, you can't base a proper decision on it. So:

- Don't make a decision unless you have to (saves restricting choices).

- Keep it under review once you've made it.

- No decision can be a decision (but watch for indecision).

Most important, though, is to *be prepared to change a decision!* (the Captain in the Dryden Accident should not have rotated twice).

Of course, by definition, the nature of most incidents means that there is no time for proper evaluation, and you have to fall back on instinct, experience or training (see *Learning*, below).

The rate of information processing is very vulnerable to fatigue and stress, and the most demands are made at the beginning and ends of a flight – the latter when you are most tired (in fact, your heart rate is most just after landing).

There are two decision-making processes that affect us, both of which really speak for themselves – *ample-time* and *time-critical*.

Ample-Time Decision Making

You start with the awareness of a situation, which means having some idea of the big picture (similar to the continual updating mentioned above). Situational awareness refers to your awareness of all relevant information. Of course, you have to know how things *should* be to recognise what's wrong! You need vigilance and continual alertness, with regard to what *may* happen on top of what *is* happening, which is kind of difficult at the end of a long day.

There are three elements to the evaluation process. Diagnosis comes first (which is more of a skill than is thought), followed by the generating of possible solutions and the assessment of any risks, which is further described below.

When evaluating a situation, you should stay as cool as possible and not let emotions cloud your decision – that is, do not let false hopes affect your thinking.

Time-Critical Decision Making

Decisions have to be made quickly, based on past experience or training – there is often no time to be creative or think up new solutions. In other words, time dictates your decision, and this is where checklists and SOPs can help, because they will be based on other peoples' experience (training is supposed to

make your actions as near to reflex actions as possible, to make way for creative thought).

Drills, as per the Ops Manual, and checklists do the same thing on a different scale. Their purpose is to provide a framework on which to base good decision-making, as well as making sure you don't forget anything. SOPs are there to provide standardisation in situations where groups are formed and dissolve with great regularity, such as flight crews.

Although a checklist doesn't contain policy, it does at least stimulate activity, since the first response of most people in an emergency is to suffer acute brainfade. Either that, or you shoot from the hip, which is equally wrong. Checklists and drills are in the Company's Ops Manual and are intended to be followed to the letter. They are not always based on the Flight Manual drills, which are required to be followed to comply with the requirements of the C of A. Whilst they have their uses, though, they can't cater for every situation, and you may have to think once in a while.

In such circumstances, it pays to have prehandled many emergencies (i.e. updating landing sites as above), but, otherwise, actions take place in two modes, the *conscious* and the *automatic*. The former can be slow and error-prone, but has more potential for being correct. The latter is largely unconscious and therefore automatic – however, it only relies on a vast database of information (or experience), and is not creative of itself – a problem that may affect inexperienced pilots.

Learning

In simple terms, this can be defined as a long-term change in behaviour based on experience, whether its other peoples' (reading, studying) or your own.

There are several behaviour patterns (or pilot performance levels) concerning this:

- *Skill-based* learning is based on practice, to become part of the "muscle memory", or *motor programs*, of your body (say when learning the piano). As such it does not require conscious monitoring, but it can lead to *environmental capture*, that is doing something because it's always done and not because it's the right thing to do (saying "3 greens", for example, without lowering the gear). You could also end up with the right skill in the wrong situation (*action slip*), meaning pulling the flap lever instead of the gear (see *Ergonomics*, below).

- *Rule-based* learning is that which follows procedures, like checklists and SOPs. It is kept in long term memory (see above), requiring the *decision channel* and *working memory* for execution - an inexperienced pilot may have a problem with this if the rules are imprecise and assume a minimum level of knowledge for them to be used properly. What usually happens when an accident occurs is that the brain goes smartly into neutral whilst everything around you goes pear-shaped. Checklists can help to bridge the gap of inactivity by giving

you something more or less correct to do whilst psyching yourself up and evaluating information ready for a decision. The US Navy, for example, trains pilots to stop in emergencies, and reset the clock on the instrument panel, which forces them to relax, or at least, not to panic.

- *Knowledge-based* learning relies on previous experience (you could look on "common sense" as the sum total of experience). It 's the sort of stuff you apply if you need to think things through, or maybe work on the *why* so the *how* becomes apparent. Inexperienced pilots are more likely to make knowledge-based mistakes, a factor that is more apparent when they are forced into knowledge-based behaviour.

Memory

Most psychologists (but by no means all!) agree that there are three types of memory:

- *Instinct*, what Jung called "race memory", gives an immediate (i.e. gut reaction) response to a stimulus. It's like something you are hard-wired for. Some psychologists call this *sensory memory*, as it provides a raw reaction to sensory input.

- *Short Term*, or *Working*, Memory, which is for data that is used and forgotten almost instantly (actually, nothing is ever forgotten, as any psychologist will tell you, but the point is that Short Term Memory is for "on the spot" work, such as fuel calculations or ATC clearances). It can only cope with about 7 items at a time, unless some tricks are used, such as grouping or association (*chunking*). Data in short term memory typically lasts about 10-20 seconds. It is affected by distraction, and is probably what Einstein was referring to when he thought that as soon as one fact was absorbed, one was discarded. Because its capacity is so limited, items must clamour for attention, which may be based on emotion, personal interest or the unusual.

Just to prove that short term memory is really limited in its holding capacity, read out the following 15 words to a few people, taking one or two seconds per word, and get them to write down as may of them afterwards as they can remember. Most people will get 7 of them, and some (around 55%) will include *sleep*, even though it wasn't there in the first place, proof that we sometimes hear what we want to hear, and that eyewitness testimony can be suspect, which is why the test was developed in the first place (by Washington University in St Louis). The words are: bed, rest, awake, tired, dream, snooze, wake, blanket, doze, slumber, snore, nap, peace, yawn, drowsy.

- *Long Term Memory*, where all our basic knowledge (e.g. memories of childhood, training, etc.) is kept – you might liken it to the unconscious. Where training is concerned, many processes can

be carried out automatically, with little thinking. Repetition is used to get information into it. It is subdivided into *semantic* memory, based on things learnt through rule-based learning (see above), and *episodic*, from specific events, for knowledge-based behaviour. Simple repetition (without meaning) is not effective for transferring items from short-term to long-term memory – there must be some sort of link, or emotion.

There are also two types of thinking:

- *Left brain*, or *logical*, involving verbal and mathematical methods. It typically uses simple deduction; for example, define what the problem is NOT until you can decide what the problem must be.

- *Right Brain*—conceptual. The artist type.

Responses

Following a decision, based on a stimulus, there is a response. However, a response due to excessive pressure is more likely to be based on insufficient data and be wrong than a more considered one, assuming time permits. Don't change a plan unnecessarily; a previously made one based on sound thinking is more likely to work than one cooked up on the spur of the moment, provided, of course that the situation is the same or similar. A correct, rather than rapid reaction is appropriate.

Response times will vary according to the complexity of the problem, or the element of expectation and hence preparedness (we are trained to expect engine failures, for example, but not locked controls, so the reaction time to the former will be less). Pushing a button as a response to a light illuminating will take about 1/5 th of a second, but add another light and button and this will increase to a second or so. An unexpected stimulus will increase reaction time to nearly 5 seconds.

There is a time delay between perceiving information and responding to it – typically 3.4 seconds. The reason we don't take this long to answer people in normal conversation is because we are anticipating what they are going to say, which leads to misinterpretation if you don't have body language to help, as with the radio.

The workload in the cockpit should be moderate; we get tired when bored, and performance is poor. Similarly, performance will be poor when you're too busy, due to swamping. There is some concern over too much automation across the Atlantic, as pilots do not have enough to do. Perhaps they should have in-flight video as well!

Human Factors

The idea behind this section is to generate countermeasures against anything that may affect your decision-making capabilities.

Physical

These are the influences that your body is subjected to.

The Environment

The conditions under which an aircraft is operated. You may be remote, in a busy area, or just bl**dy cold.

You can feel cooler because moisture is evaporating from your skin at an advanced rate in dry air - humidity would normally need to be 60% at 18°C for comfort. As far as temperature goes, the human body operates comfortably in a range between 18-24°C.

Time

Pressure from customers and employers to keep to deadlines.

Air Quality

Not only can haze or mist reduce visibility, but it can also irritating, or smelly, or deadly (carbon monoxide).

CO is a colourless, odourless gas which has a half life of about six hours at sea level pressures, so a quarter is present after 12 hours. It typically gets into the cockpit from faulty exhausts, and also comes about where something is burning without an adequate air supply, or where combustion is incomplete. Characteristic symptoms of carbon monoxide poisoning include cherry red lips.

Flicker

This occurs when light is interrupted by propeller or rotor blades. It can cause anything from mild discomfort to fatigue, and even convulsions or unconsciousness. Flicker certainly modifies certain neuro-physiological processes – 3-30 a second appears to be a critical range, while 6-8 will diminish your depth perception (the Nazis set their searchlights to flicker, to get up the nose of bomber pilots). Hangovers make you particularly susceptible.

Noise, Vibration & Turbulence

Prolonged amounts of any of these is fatiguing and annoying. Noise is particularly prevalent in helicopters, especially with the doors off. Vibration at the right frequency will cause back pain.

Ergonomics

Under this comes cockpit design and automated systems. Here's an illustration of how bad design can be the start of an *event chain*:

A relatively inexperienced RAF Phantom (F4) pilot had a complete electrics failure, as if being over the North Sea at night in winter wasn't stressful enough. For whatever reason, he needed to operate the Ram Air Turbine, but he deployed the flap instead, as the levers were close together.

Of course, doing that at 420 knots made the flaps fall off the back, and the hydraulic fluid followed. Mucking around with the generators got the lights back on, and he headed for RAF Coningsby, with no brakes. Unfortunately, the hook bounced over the top of the arrester wire, so he used full afterburner to go around in a strong crosswind, but headed towards the grass instead. The pilot and navigator both ejected, leaving the machine to accelerate through 200 knots, across the airfield at ground level.

Meanwhile, the Station Commander was giving a dinner party for the local mayor in the Mess, and the guests had just come out on the steps (not far from the runway), just in time to watch the Phantom come past on the afterburner, with two ejections. The mayor's wife was just thanking him for the firework display as it went through a ditch, lost its undercarriage and fell to bits in a field.

The Fire Section had by this time sent three (brand new) appliances after it without any hope of catching up, but they tried anyway. The first one wrote itself off in a ditch because it was going too fast, the driver of the second suddenly put the brakes on because he realised there had been an ejection and that he might run over a pilot on the runway, at which point the number three appliance smashed into the back of him.

We are in a similar situation – how many times have you jumped into the cockpit of a different machine, to find the switches you need in a totally different place? This doesn't help you if you rely on previous experience to find what you need (in emergencies you tend to fall back to previous training), so the trick here is to know what you need at all times, and take the time to find out where it is (and *read the switches*).

Physiological (The Body)

The human body is wonderful, but only up to a point. It has limitations that affect your ability to fly efficiently, as your senses don't always tell you the truth, which is why you need extensive training to fly on instruments, as you have to unlearn so much. The classic example is the "leans", where you think you're performing a particular manoeuvre, but your instruments tell you otherwise.

Why do you need to learn about the body? Well, parts of it are used to get the information you need to make decisions with. And, of course, if it isn't working properly, you can't process the information or implement any action based on it.

G Tolerance

The body can only cope with certain amounts of *G-force*, from the effects of acceleration. When there is none, you are subject to 1G.

Negative G acts upwards and can increase the blood flow to the head, leading to *red out*, facial pain and slowing down of the heart. In addition, your lower eyelids close at –3G. *Positive G* is more normal, but will drain the blood, with the obvious consequences, including loss of vision, called *grey out*, at +3.5G This could end up as *black out* and unconsciousness at +6G. Both are affected by hyperventilation, hypoxia, heat, hypoglycaemia, smoking and alcohol, all of which are discussed more fully below.

The *valsalva manoeuvre* can be used to help cope with high G (close your throat and hold your nose), or you could use a pressure suit.

Impact-wise, the body can tolerate 25G vertically and 45G horizontally - if you don't wear shoulder straps, tolerance to forward deceleration reduces to below 25G, and you will jackknife over your lapstrap with your head hitting whatever is in front of it at 12 times the speed it is coming the other way.

Body Mass Index (BMI)

This is calculated by your weight in kg divided by your height squared, in metres squared.

If it is over 30, you are obese. This could lead to diabetes and heart disease, and reduce your ability to cope with hypoxia, decompression sickness and G tolerance.

The Central Nervous System

This consists of the brain and spinal cord, though, for exam purposes, it also includes the visual and aural systems (eyes and ears), proprioceptive system and chemosensory system (smell).

The *autonomic* nervous system regulates vital functions over which you have no conscious control, like heartbeat and breathing (unless you're a high-grade Tibetan monk, of course).

Although the brain is only 2% of the body mass, it takes up to 20% of the volume of each heartbeat – its blood supply needs to be continuous, as the brain cannot store oxygen.

Eyes

Vision is your primary (and most dependable) source of information. It gets harder with age to distinguish moving objects; between the ages of 40 and 65, this ability diminishes by up to 50%. This is only one of the limitations of sight, and we need to examine the eye to see how you overcome them.

The eye is nearly round, and its rotation in its socket is controlled by external muscles:

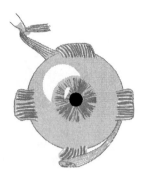

It has three coatings; the *sclerotic*, which is transparent at the front; the *choroid*, which lines the sclerotic and contains tiny blood vessels, and the *retina*, which is the light sensitive bit that detects electromagnetic waves of the frequency of light, and converts them to electrical signals that are interpreted by the brain, and which is sensitive to hypoxia.

This means you see with the brain as well, giving a difference between *seeing* and *perceiving*, discussed below. This is because the eye's optical quality is actually very poor (in fact, you would get better results from a pinhole camera), hence the need for the brain, which can actually modify what you see, based on experience, and so is reliant on expectations. If the brain fills in the gaps wrongly, you get visual illusions.

The eye can, however react quickly to changes in light conditions, although it is slower to adapt from light to dark because a chemical (*visual purple*) needs to be created.

The *optic nerve* carries signals from the eye to the brain, the *lens* focuses light on to the retina, the *iris* controls the diameter of the *pupil* (the black bit in the middle where light gets through), and the *cornea* refracts light onto the lens. The area of sharp vision is actually very small (at 4 feet the size of a small coin), because of the relatively small size of the *fovea*. 5°away from the foveal axis, it reduces by a quarter, and one-twentieth at 20° away. You should be able to see another aircraft directly at 7 miles, or 2.5 miles if it was 45° off – at 60° it's down to half a mile! The reason why you are taught to scan is because the eye needs to latch on to something, difficult to do with a clear blue sky. With an empty field of vision, your eyes will focus at short distances, about 1-2 metres ahead, and miss objects further away.

Peripheral vision also has a different neurological route to the brain. To prove this, stand up, cover one eye and make a fist with the other, to cover up the good eye's central vision. Now stand on one foot. Now do it again, with your forefinger and thumb making a very small circle where the fist was – in other words, you are using your central vision to look through the circle. Standing on one foot is now more difficult – and this is what we look at our instruments with!

Accommodation is the change in refraction of light as it enters the eye, caused by altering the curvature of the lens.

The major causes of defective vision are:

- *Hypermetropia* – where the eyeball is too short, and images focus behind the retina (log sight). Requires a convex lens.

- *Myopia* – where the eyeball is too long, and images focus in front of the retina (short sight). Needs a concave lens.

- *Presbyopia* – the lens hardens, leading to hypermetropia (comes with old age).

- *Cataracts* – the lens becomes opaque.

- *Glaucoma* – pressure inside the eyeball.

- *Astigmatism* – unequal curvature of the cornea or lens.

Laser surgery, by the way, only corrects one aspect of vision, that is, long or short sight, but not both.

Rods and Cones

The retina consists of light sensors (actually, neurons) which are called *rods* and *cones*, because of the way they are shaped. Each is more efficient than the other in different kinds of light. The point where the optic nerve joins the retina is mostly populated with cones, which work best in daylight and become less effective at night, or where oxygen levels are reduced (which is significant for smokers, whose blood has less oxygen carrying capacity), so you get a blind spot in the direct field of vision, which is why you see things more clearly at night if you look slightly to the side of what you want to look at. Rods cannot distinguish colours, either, which is why things at night seem to be in varying shades of grey (you see colours simply because the vibrations they give out are strong enough to wake the cones up).

The more your iris is open, the less *depth of field* you have, so in darkness it is difficult to see beyond or before a certain distance, and you may require glasses to help (the depth of field in photography is an area between the camera and the subject in which everything is in focus. The wider the aperture, or iris, the shorter this distance is, and *vice versa)*.

The retina contains enormous amounts of vitamin A, which is necessary for adapting between light and darkness. Too little vitamin A could therefore result in *night blindness*. The changeover from light to dark takes about 20-30 minutes and should always be allowed for when night flying (actually, the cones take 7 minutes, and the retina can take up to 45 minutes).

The greatest visual acuity (that is, the ability to see small detail) is obtained by cones in the *fovea*, in the central part of the retina, so you must look directly at an object to see it best. At night, look slightly to one side, because the rods that are sensitive to lower levels are outside the fovea, at the peripheral of the retina.

With distance, objects on the retina become smaller.

Refraction
The transparent part of the sclerotic is known as the *cornea*, behind which is the *lens*, whose purpose is to bend light rays inwards, so they focus on the retina. If this happens in front of it, *short sightedness* results. You get long sightedness with the point of focus behind the retina. Both conditions cause blurred vision, correctable by glasses, that vary the refraction of the light waves until they focus in the proper place.

Blurred vision can also be caused by stress, causing nervous tension, and excessive eye muscle activity leading to eyestrain. Just relaxing often helps this condition.

70% of light is refracted by the cornea, and 35% by the lens.

Optical illusions
Searching for an object in a swimming pool is difficult, because the light rays bend as they pass the surface and the object appears to be displaced. Similarly, rain on a windscreen at night gives the impression that objects are further away. A good fixed wing example of an optical illusion is a wider runway tending to make you think the ground is nearer than it actually is; a narrow runway delays your reactions, possibly leading to a late flare and early touchdown. Any object (like a runway) that you think is smaller than it actually is, will

also appear to be nearer, and vice versa. If the object is brighter than its surroundings (a well-lit runway), you will think you are higher than you are, so on an approach, you might start early and be lower than you should. In haze, objects appear to be further away.

An approach to a downsloping runway should be started higher, with a steeper angle, and one to an upsloping runway should be started lower, at a shallower angle.

When mountain flying, it's often difficult to fly straight and level because the sloping ground around affects your judgment. Similarly, you can't judge your height when landing on a peak. Even going to the cinema is an optical illusion; still frames are shown so quickly it looks as if movement is taking place—the switching is done in the brain.

Vectional illusions are caused by movement, as when sitting in a railway carriage and wondering whether it's the train next to you or the one you're in that is moving. The *autokinetic* effect is the illusion that an object is moving, where it is actually your eye that is moving. Distant objects become less colourful and less distinct.

A high speed aircraft approaching head on will grow the most in size very rapidly in the last moments, so it's possible for an approaching aircraft to be hidden by a bug on the windscreen for a high proportion of its approach time (you might only see it in the last few seconds). Lack of relative movement makes an object more difficult to detect.

Ears

These are important because an auditory stimulus is the one most often attended to. How many times do you answer the phone when you're busy, even though you've ignored everything else for hours?

Sound waves make the *eardrum* vibrate, and the vibrations are transmitted by a chain of linked bones known as the *hammer, anvil* and *stirrup* to the inner ear, which is full of fluid. There are thousands of fibres of different lengths within the inner ear which vibrate in sympathy according to different frequencies.

As some of the fibres get damaged (through too severe vibration), the ability to hear that frequency goes (they do not regenerate). *Presbyacusis* is hearing loss with age, where the high tones go first. *Noise Induced Hearing Loss*, or NIHL, occurs through prolonged exposure to loud noise, usually 90 db and above. The fibres are linked to the brain and, as with sight, it is now, when the signal reaches the brain, that we "hear".

The *semicircular canals* are what we use to keep balanced:

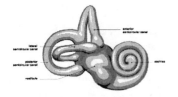

They use the fluid in the inner ear, which acts against sensors to send electrical signals to the brain so you can tell which way up you are (they detect angular acceleration). The leans happen because your semicircular canals get used to a particular sustained motion in a very short time. If you start a turn and keep it going, your canals will think this is normal, because they lag, or are slow to respond. When you straighten up, they will try to tell you you're turning, where you're actually flying straight and level. Your natural inclination is to obey your senses, but your instruments are there as a cross-reference. In fact, the whole point of Instrument Training is to overcome your dependence on your senses. Particularly dangerous is recovering from a spin of 2-3 turns, where you think you are actually turning the opposite way and enter another spin when you try to correct it. Eventually an extreme nose-up condition results, which turns into an extreme nose-down attitude and a tight *graveyard spiral* before entering Terrain Impact Mode.

The *coriolis illusion* is easily demonstrated with a revolving chair – sit in one, and get someone to spin it while you have your chin on your breast. When you raise your head sharply, you will find yourself on the floor inside two seconds. This has obvious parallels with flying, so make all your head movements as gently as possible, especially when making turns or other manoeuvres in IMC.

You can get problems from colds, etc as well, particularly a spinning sensation caused by a sudden difference in pressure between the inner portions of each ear.

The *Eustachian Tubes* are canals that connect the throat with the middle ear; their purpose is to equalise air pressure. When you swallow, the tubes open, allowing air to enter, which is why swallowing helps to clear the ears when changing altitude. Blocked Eustachian tubes can be responsible for split eardrums, due to the inability to equalise pressure. Since the eardrum takes around 6 weeks to heal, the best solution is not to go flying with a cold, but commercial pressures don't always allow this. If you have to, make sure you use a decongestant with no side effects.

The audible range of the human ear is 20 Hz to 20 KHz.

Blocked Sinuses
Although associated with the nose, the sinuses are actually hollow spaces or cavities inside the head surrounding the base of the nose and the eye sockets. Amongst other things, they act as sound boxes for the voice. Being hollow, they provide structural strength whilst keeping the head light; there are normally between 15-20.

Blockages arise from fluid that can't escape through the narrow passages—pain results from fluid pressure. Blocked sinuses can also be responsible for severe headaches.

Deafness
This can arise from many causes; in aviation, high-tone deafness from sustained exposure to jet engines is very common. Hearing actually depends on the proper working of the *eighth cranial nerve*, which carries signals from the inner ear to the brain. Obviously, if this gets damaged, deafness results. The nerve doesn't have to be severed, though; deterioration will occur if you don't get enough Vitamin B-Complex (deafness is a symptom of beriberi or pellagra, for example, from Vitamin B deficiency).

You can recover from some deafness, such as that caused by illness, but not that caused by damage to the fibres in the fluid.

Disorientation
This refers to a loss of your bearings in relation to position or movement. The "leans" is the classic case, already mentioned. To combat them, close your eyes and shake your head vigorously from side to side for a couple of seconds, which will topple the semi-circular canals. Motion Sickness usually happens because of a mismatch between sight, feel and the semicircular canals, giving unfamiliar real or apparent motion (e. g. the leans). Medication can have unwelcome side effects, particularly on performance, which are normally not acceptable for flight crews.

During acceleration, it's possible to get the impression of pitching up, making you want to push the nose down (it's more pronounced at night going into a black hole from a well-lit area). You get a pitch-down illusion from deceleration. The danger here is that lowering the gear or flaps causes the machine to slow down, which makes you think you are pitching down and want to bring the nose up, which could cause a stall at the wrong moment on approach.

Respiratory System
Consists of the lungs, oronasal passage, pharynx, larynx, trachea, bronchi, bronchioles and alveoli:

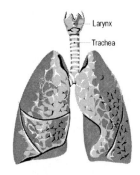

— Larynx

— Trachea

Air is drawn into the lungs, from where oxygen is diffused into the haemoglobin in the blood under pressure, which carries it to the tissues of the body, especially the brain, the most sensitive to lack of it. Blood is pumped around by the heart. Waste products in the form of carbon dioxide go the other way, via plasma to the lungs (it is the carbon dioxide level in the blood that regulates respiration, which is monitored by chemical receptors in the brain that are very sensitive to CO_2). The diffusion of oxygen into the blood depends on *partial pressure* (that is, the pressure in proportion to the amount of an individual gas in the mix), so as this falls, oxygen assimilation is impaired (although the air gets thinner, the ratio of gases

remains the same). Even if you increase the proportion of oxygen to 100% as you climb, there is an altitude (around 33,700 feet) where the pressure is so low that the partial pressure is actually less than that at sea level, so just having oxygen is not enough.

From 0-10,000 ft you can survive on normal air; above this, an increasing amount of oxygen relative to the other ingredients is required, up to 33,700 feet, at which point you require pure oxygen to survive (breathing 100% oxygen at that height is the same as breathing air at sea level. At 40,000' it is the equivalent of breathing air at 10,000 feet). Above 40,000 the oxygen needs pressure (also, exposure to O_3 becomes significant). Having said all that, your learning ability can be compromised as low as 6000 feet (*Source*: RAF).

The normal rate of breathing is around 18 times a minute, exchanging .35-.65 litres of air.

Oxygen
Pure oxygen is a colourless, tasteless, odourless and non-combustible gas that takes up about 21% of the air we breathe. Although it doesn't burn itself, it does support combustion, which is why we need it, because the body turns food into heat. As we can't store oxygen, we survive from breath to breath.

How much you use depends on your physical activity and/or mental stress—for example, you need 4 times more for walking than sitting quietly. The proportion of oxygen to air (21%) actually remains constant, but as the air gets less dense, each lungful contains less oxygen in proportion (that is, the *partial pressure* becomes less), which is why high altitude flight requires extra supplies. Nothing more is required below 5000 feet, as 95% of what you would find on the ground can be expected there. However, at over 8000 feet, you may find measurable changes in blood pressure and respiration, although healthy individuals should perform satisfactorily.

Lack of oxygen leads to...

Hypoxia

A condition where you don't have enough oxygen in the tissues, resulting from inefficient transfer of it into the blood, but anaemia can produce the same effect, as can alcohol (there are actually several types of hypoxia, but we won't bother with that here). In other words, there may really be too little oxygen, or you don't have enough blood to carry what you need around the body—you may have donated some, or have an ulcer. You might also be a smoker, with your haemoglobin blocked by carbon monoxide (*anaemic hypoxia*). A blockage of 5-8%, typical for a heavy smoker, gives you the equivalent altitude of 5-7000 feet before you even start!

The effects of hypoxia are similar to alcohol but the classic signs are:

- *Personality changes.* You get jolly, aggressive and less inhibited.

- *Judgement changes.* Your abilities are impaired; you think you are capable of anything and have much less self-criticism.

- *Muscle movement.* Becomes sluggish, not in tune with your mind.

- *Short-term memory loss.* This leads to reliance on training, or procedures established in long-term memory.

- *Sensory loss.* Blindness occurs (colour first), then touch, orientation and hearing are affected.

- *Loss of consciousness.* You get confused first, then semi-conscious, then unconscious.

- *Blueness.*

The above are *subjective* signs, in that they need to be recognised by the person actually suffering from hypoxia, who is actually in the wrong state to recognise anything. External observers may notice some of them, but especially lips and fingertips turning blue and possible hyperventilation (see below) as the victim tries to get more oxygen.

All are aggravated by:

- *Altitude.* Less oxygen available, and less pressure to keep it there.

- *Time.* The more exposure, the greater the effect.

- *Exercise.* Increases energy usage and hence oxygen requirement.

- *Cold.* Increases energy usage and hence oxygen requirement.

- *Illness.* Increases energy usage and hence oxygen requirement.

- *Fatigue.* Symptoms arise earlier.

- *Drugs or alcohol.* Reduced tolerance.

- *Smoking.* Carbon monoxide binds to blood cells better than oxygen.

The *times of useful consciousness* (that is, from the interruption of the oxygen supply to when you can do nothing about it) are actually quite short:

Height	Time
18,000'	20-30 mins
22,000'	5-10 mins
25,000'	2-3 mins
28,000'	1-1 ½ mins
30,000'	45-75 secs
35,000'	25-35 secs
45,000'	12-20 secs

Oxygen Requirements

The oxygen to be carried, and the people to whom masks should be made available, varies with altitude, rate of descent and MSA. The latter two are dependent on each other, in that it's no good having a good rate of descent if the MSA stops you. It may well be that, although you're flying at a level that requires fewer masks, the MSA may demand that you equip everybody.

Preflight stuff includes ensuring that oxygen masks are accessible for the crew, and that passengers are aware of where their own masks are. Check the security of the circular dilution valve filter (a foam disc) on all of them, together with the pressure. Beards will naturally reduce their efficiency. Briefings should include the importance of not smoking and monitoring the flow indicator. All NO SMOKING signs should be on when using it.

If you know you will need oxygen at night, it's best to start using it from takeoff.

There are three types of oxygen supply, *continuous flow, diluter demand* and *pressure demand.* Refer to the *Air Law* chapter for legal requirements.

Hyperventilation

This is simply overbreathing, where too much oxygen causes carbon dioxide to be washed out of the bloodstream, which then gets too alkaline (oxygen is actually quite corrosive – it belongs to the same chemical family as chlorine and fluorine, so too much is toxic). Unconsciousness slows the breathing down so that the CO_2 balance is restored, but falling asleep is not often practical! The usual cause is worry, fright or sudden shock, but hypoxia can be a factor.

Symptoms include:

- Dizziness

- Pins and needles, tingling

- Blurred sight

- Hot/Cold feelings

- Anxiety

- Impaired performance

- Loss of consciousness

Pressure Changes

Aside from oxygen, the body contains gases of varying descriptions in many places; some occur naturally, and some are created by the body's normal working processes. The problem is that these gases expand and contract as the aircraft climbs and descends. Some need a way out, and some need a way back as well.

- *Gas in the ears* is normally vented via the Eustachian tubes. If these are blocked (say with a cold), the pressure on either side of the eardrum is not balanced, which could lead (at the very least) to considerable pain, and (at worst) a ruptured eardrum.

- *Sinus cavities* are also vulnerable to imbalances of pressure, and are affected in the same way as eardrums.

- *Gas in the gut* can be vented from both ends.

- *Teeth* may have small pockets of air in them, if filled, together with the gums. Although dentists nowadays are aware of people flying, and pack fillings properly, the general public don't fly every day, as you do, so it's best to be sure. High altitude balloonists actually take their fillings out.

Motion Sickness

This is caused by a mismatch between the information sent to the brain by the eyes and ears.

Accelerating from straight and level may give you the impression of pitching up, because the sensors in the inner ear perceive the body weight as going rearwards and downwards. As the most dependable source of sensory information is your eyes, *believe your instruments.*

Decompression Sickness

Where pressures are low, nitrogen in the blood comes out of solution (typically above 18,000', but more so at 25,000'). Bubbles can form, and are especially painful in the joints (e.g. the bends, for the joints, the *creeps* (skin), *chokes* (lungs) and the *staggers* (brain).

Unfortunately, these bubbles do not redissolve on descent, so if you are affected you may need to go into a decompression chamber. For this reason, diving before flight should be avoided, as extra nitrogen is absorbed while breathing pressurised gas and will dissolve out as you surface again. Don't fly for 12 hours if you have been underwater with compressed air, and 24 hours if you've been below 30 feet.

Circulatory System

Made up of the heart, arteries, arterioles, capillaries, venules, veins and blood.

The Heart

This item does not rest in the same way as other muscles – instead, it take a mini-rest for a microsecond or two in between beats (a normal pulse rate is around 70 beats per minute).

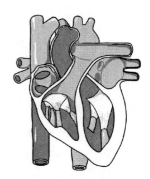

The *systolic* blood pressure is the peak pressure as blood is pumped from the heart to the aorta. The *diastolic* pressure is the lowest, produced when resting between beats. Normal blood pressure lies between 110-145 mmHg (systolic) and 70-90 mmHg (diastolic) – you might see a figure of 120/80 (120 over 80).

Blood

55% *plasma*, for transporting CO_2, nutrients and hormones, and 45% *cells*. Red cells transport oxygen via haemoglobin, and white cells (*leukocytes*) fight infection. *Platelets* are for clotting blood.

Heart Disease

Heart disease can be grouped into 3 categories:

- *Hypertensive*— from high blood pressure, working the heart harder so it gets enlarged (anxiety, etc.).

- *Coronary,* or *Arteriosclerotic*— hardening of the arteries through excessive calcium, or cholesterol, which again makes the heart work harder (bad diet).

- *Valvular* or *rheumatic*— where valves are unable to open or close properly, allowing back pressure to build up (old age).

Adrenaline increases the speed and force of the heart beat.

To reduce the risks of heart disease, double your resting pulse for at least 20 minutes 3 times a week.

Fatigue

It's not so much the amount of sleep you get, but when you get it that counts, so fatigue is just as likely to result from badly planned sequences of work and rest. A surprising amount (over 300) of bodily functions depend on the cycle of day and night—we have an internal rhythm, which is modified by such things. You naturally feel best when they're all in concert, but the slippery slope starts when they get out of line. The best known desynchronisation is jet lag, but it also happens when you try to work nights and sleep during the day— bright light can fool your body into thinking it's day when it's not. One day for each time zone crossed is required before sleep and waking cycles get in tune with the new location, and total internal synchronisation takes longer (kidneys may need up to 25 days).

Even the type of time zone change can matter—6 hours westward requires (for most people) about four days to adjust—try 7 for going the other way! This Eastward flying compresses the body's rhythm and does more damage than the expanded days going west; North-South travel appears to do no harm.

Symptoms of jet lag are tiredness, faulty judgement, decreased motivation and recent memory loss. They're aggravated by alcohol, smoking, high-altitude flight, overeating and depression, as found in a normal pilot's lifestyle.

In view of this, commercial pilots have a maximum working day that is intended to ensure they are rested enough to fly properly.

The two types of fatigue are *acute* and *chronic*, the former being short-term, or more intense, and the latter arising from more long-term effects, like many episodes of acute fatigue.

Foods low in carbohydrate or high in protein help fight fatigue, especially "healthy" ones, like fruit or yoghurt, or cereals, especially granola. Coffee, of course, contains caffeine, which keeps you awake (as does tea), but too much can lead to headaches and upset stomachs. Be aware that it takes about 5 minutes for caffeine to kick in, and people who drink unleaded coffee (decaffeinated) still report the unpleasant side effects (the process that removes the caffeine is allegedly just as harmful in different ways). Caffeine has a half-life of about 3 hours, and although it might not stop you getting to sleep, it will affect the quality of that sleep.

Sleep is actually a state of altered consciousness, in which you don't lose awareness of the external world, as any mother will tell you. It is part of a daily cycle which is actually 25 hours long – that is, the sleeping and waking rhythm is about an hour longer than the normal day of 24 hours, which itself is a mean figure anyway (Moore-Ede, Sulzman & Fuller, 1982). This is why flying West is easier on the system than flying East – the body's rhythm is extended in the right direction.

It's worth noting that a shave is about equal to 20 minutes of sleep, in terms of refreshing you, and washing your face or brushing your teeth are also good ideas, as is moving around for 5-10 minutes.

Types of Sleep

In a typical cycle of 90 minutes, you descend through 4 stages of NREM (*non-REM*) sleep, each deeper than the other, and return, when you enter the REM state, where most dreams occur. This is an *ultradian rhythm*, which comes from the Latin *ultra dies*, or "faster than a day". REM sleep (*Rapid Eye Movement*) refreshes the mind, and *Slow Wave* (NREM) sleep refreshes the body. You are more refreshed if you wake up during the former. Stages 3 and 4 are known as *slow wave* sleep, after the patterns on an EEG, and you are more groggy if you wake up in that period. You might go through 4 or 5 episodes of REM sleep a night.

It is possible that the nature of the fatigue determines the sleep stages required. For example, extra slow-wave sleep is required after physical activity.

1 hour of sleep is equal to 2 hours of activity, and you can accumulate up to 8 hours on a credit basis. Alcohol interferes with the latter period of sleep with its diuretic action (you need a pee) – repeated use disturbs sleep patterns on a long term basis, to give you insomnia.

Clinical Insomnia is being unable to sleep under normal conditions. *Situational insomnia* arises out of the circumstances, like sleeping in a strange bed or time zone (*circadian desynchronisation*). Although insomniacs may think they don't sleep at all, they actually spend their time in stages 1 and 2.

Diet
The body's main fuel is glucose, which can either be converted from different types of food, or eaten directly.

Levels of glucose are regulated by the pancreas, which secretes insulin to reduce high levels of blood sugar by converting it into fat.

Hypoglycaemia
The most common problem (in the normal pilot's lifestyle, anyway) is *low* blood sugar, caused by missed meals and the like. Although you may think it's better to have the wrong food than no food, be careful when it comes to eating choccy bars in lieu of lunch, which will cause your blood sugar levels to rise so rapidly that too much insulin is released to compensate, which drives your blood sugar levels to a lower state than they were before—known in the trade as "rebound hypoglycaemia". Apart from eating "real food", you will minimise the risks of this if you eat small snacks frequently instead of heavy meals after long periods with nothing to eat. Complex carbohydrates are best, in the shape of pasta, etc.

Hypoglycaemia is fair enough in the short term, but long-term can be a *disease*. Although not life threatening, hypoglycaemia is a forerunner of many worse diseases and should be looked

at. The important thing to watch appears to be the suddenness of any fall in blood sugar, and a big one can often trigger a heart attack. A high protein diet will tend to even things out, as protein helps the absorption of fat, which is inhibited if too much insulin is about (also check out *Food combining*, where proteins and starches are not mixed due to chemical incompatibility).

Warning signs include shakiness, sweatiness, irritability or anxiety, difficulty in speaking, headache, weakness, numbness or tingling around the lips, inability to think straight, palpitations and hunger.

At its worst, hypoglycaemia could result in coma, but you could also get seizure and fainting. Eat more if you exercise more.

Hyperglycaemia
This is the opposite, being an excess of blood sugar. Symptoms would include tiredness, increased appetite and thirst, frequent urination, dry skin, flu-like aches, headaches, blurred vision and nausea. This condition causes dehydration, so always have fluids around to help you. Also, decrease stress.

Alcohol
Whilst nobody should object to you taking a drink or two the evening before a flight, you should remember that it can take over 3 days for alcohol to clear the system (it remains in the inner ear for longest). Within 24 hours before a

planned departure, you should not drink alcohol at all; certainly not on standby. The maximum blood level is officially .2 mg per ml, a quarter of the driving limit in UK, but it's not only the alcohol that causes problems – the after-effects do as well, like the hangover, fatigue, dehydration, loss of blood sugar and toxins caused by metabolisation.

Although it appears otherwise, alcohol is not a stimulant, but an anaesthetic, which puts to sleep those parts of the brain that deal with inhibitions - the problem is that these areas also cover judgement, comprehension and attention to detail. In fact, the effects of alcohol are the same as hypoxia, dealt with elsewhere, in that it prevents brain cells from using available oxygen. One significant effect of hypoxia in this context is the resulting inability to tell that something is wrong.

It takes the liver about 1 hour to eliminate 1 unit of alcohol from the blood (officially, alcohol leaves the body at a rate of 15 milligrams per 100 ml of blood per hour). 1 unit is considered to be the same as 1 measure of spirit, a glass of wine or half a pint of beer. The number of units per week beyond which physical damage is likely is 21 for men and 14 for women.

As far as passengers are concerned, although they get cabin service, persons under the affluence of incohol or drugs, of unsound mind or having the potential to cause trouble should not be allowed on board—certainly, no person should be drunk on any aircraft (people aren't generally aware that one drink at 6000 feet is the same as two at sea level). This is not being a spoilsport—drunks don't react properly in emergencies and could actually be dangerous to other people (which is why I always get an aisle seat – I don't have to get round people in the way). Therefore, it's not just for their own good, but that of others as well. If you need to get rid of obstreperous passengers (long word number 1), you can always quote the regulations at them (or even use sarcasm), but don't forget to fill in an Occurrence Report.

Medications

Although the symptoms of colds and sore throats, etc. are bad enough on the ground, they may actually become dangerous in flight by either distracting or harming you by getting more serious with height (such as bursting your eardrums, or worse). If you're under treatment for anything, including surgery, not only should you not fly, but you should also check there will be no adverse effects on your physical or mental ability, as many preparations combine chemicals, and the mixture could make quite a cocktail. No drugs or alcohol should be taken within a few hours of each other, as even fairly widely accepted stuff such as aspirin can have unpredictable effects, especially in relation to Hypoxia (it's as well to keep away from the office, too— nobody else will want what you've got). Particular ones to avoid are antibiotics (penicillin, tetracyclines), tranquilisers, antidepressants, sedatives, stimulants (caffeine, amphetamines), anti-histamines and anything for relieving high blood pressure, and, of course, anything not actually prescribed. Naturally, you've got to be certifiable if you fly having used marijuana, or worse.

Blood Donations
Pilots generally are discouraged from giving blood (or plasma) when actively flying, and some dental anaesthetics can cause problems for up to 24 hours or more, as can anything to do with immunisation. If you do give blood, try to leave a gap of 24 hours, including bone marrow donations. Although blood volume is restored in a very short time, and for most donors there are no noticeable after-effects, there is still a slight risk of faintness or loss of consciousness. After a general anaesthetic, check with the doctor.

Food poisoning can also be a problem, and not just for passengers—the standard precaution (like in *A irplane!*) is to select different items from the rest of the crew, even in the hotel.

Don't forget to inform the authorities (in writing) of illnesses, personal injuries or presumed pregnancies that incapacitate you for more than 20 days (you can fly up to the 30th week of pregnancy in Canada, if your doctor agrees, but wait for 4-6 weeks afterwards before flying again). Pilots involved in accidents should be medically examined before flying again.

Radiation
At high altitudes, the main source is energetic particles from outside the solar system. Excessive exposure to it may lead to cancer.

Exercise
Physical exercise strengthens the heart and enhances the blood flow,

which helps when you get hypoxia. More of this below.

Psychological Factors
That which influences, or ends to influence, the mind or emotions. We are concerned with them, because they can influence the way we interpret information on which we base decisions.

The thing is, with both optical and aural stimuli (discussed above), the processing is done in the brain, which uses past experience to interpret what it senses – it therefore has expectations. Even though the transmissions are bad, we often fill in gaps in messages from ATC because we have heard them before and know what to expect. Similarly, when sitting in a train, we think we are moving, when all the time it is the other train. Because the ears are also our organs of balance, we can suffer illusions of movement.

Stress
Flying is stressful, there's no doubt about that, but should stress be a problem? It's arguable that a little is good for you; it stops you slowing down and keeps you on your toes; this is the sort associated with success. Excessive stress, on the other hand, in the form of pressure (that is, stress without respite) can lead to fatigue, anxiety and inability to cope, and is associated with frustration or failure.

Stress and preoccupation have their effects; a PA31 pilot was doing a cargo flight with three scheduled stops, but he did not refuel or even shut down at any of them, so both engines stopped after the last delivery. He was anxious to get home as his wife was in hospital.

Fight and flight responses are bodily changes that prepare it for action – adrenaline starts to pump and many other changes take place as well, including a rise of sugar in the blood. When under stress you may well revert to former training – watch out for those levers in the wrong place on the new machine!

What is Excessive Stress?

Anything that has a sufficiently strong influence to take your mind off the job in hand, or to make you concentrate less well on it. Not only are you not doing your job properly, but subconsciously feel guilty about it, too, which is enough to set up a little stress all of its own. We all like to feel we are doing the best we can possibly do, and it disturbs our self-image to feel that we're not. Consequently we get angry at ourselves for being in such a position, which increases the stress, which further takes us away from the job, and so it circulates.

Common situations causing this are:

- Grief
- Divorce
- Financial worries
- Working conditions
- Management pressure
- Pride
- Anger
- Get-home-it is

- Motivation
- Doubts (about abilities, etc)
- Timetable
- Passengers expectations and timetables

Particular to aviation is trying to beat the weather, not having enough fuel and flying when ill.

All of these lead to anxiety, which is really based on fear, if you think about it. As anxiety can cause stress, you get a circulating problem. You could probably think of more. People have their own ways of dealing with stress, so what works for one does not necessarily for someone else. This is possibly because of the evaluation of the stress that that particular person has, i.e. whether they feel they can cope and their perception of the problem.

It is *perception* of demands and abilities, rather than actual problems that affect the individual. If you *feel* you are capable, your stress level will be relatively low.

Symptoms of stress include:

- Detachment from the situation
- Failure to perceive time
- Fixation of attention
- Personality changes
- Voice pitch changes
- Desire for isolation
- Reduced cognitive ability

- Poor emotional self-control

- Unsafe cavalier attitude

Coping With Stress

You can either adjust to the situation, or change the situation itself. The willingness to recognise stress and to do something about it must be there; for example, if you don't admit there's a problem at home, there's not much you can do! It is not weakness to admit you have a problem—rather, it shows lack of judgement if you do otherwise.

Stress Management

The point of stress management programs is to help you see and recognise stress, not to cure it. You can help, however, in these ways:

- Reduce the load

- Reduce self-medication

- Exercise

- Proper diet.

- Keep a positive outlook.

Organisational Factors

That is, the people we work for, or with. An organisation is a structure within which people work together in an organised and coordinated way to achieve certain goals.

The culture of the organisation can have a significant bearing on how people perform within it. Their goals may conflict, for example, resources may be insufficient, as may planning or supervision. We all know about pressures, commercial or otherwise.

Communication

Defined as the ability to put your ideas into someone's head and be sure of success, or to exchange information without it being changed. Or both. Unfortunately, even under ideal conditions, only about 30% is retained, due to inattention, misinterpretation, expectations and emotions. Your team needs to know what you want done, and require feedback as to progress and satisfaction of your expectations. This could be through the spoken word or body language.

7% of all communication is accomplished verbally, 38% by unconscious signals, such as tone of voice, and the remainder (55%) by non-verbal means, i.e. body language, as mentioned above. *Barriers to communication* include reluctance to ask questions, the influence of authority, and difficulty in listening. You, therefore, have to put people at their ease and make them think they can talk to you or ask questions.

So, communication is the exchange of thoughts, messages or information by various means, including speech. The elements of the process are the sender, the message, the receiver and the feedback.

The perceptions and background of people at either end may influence what is sent – (a person at one end of a radio transmission might receive "send three and fourpence, we are going to a dance" instead of "send reinforcements, we are going to advance".

You could emphasise each word in turn of the following sentence and have a different meaning every time:

I never said your dog was ugly

Effective communication demands five skills:

- *Seeking information* - good decisions are based on good information, so we need it to do our jobs effectively – particularly with reference to finding out what the customer *actually* wants.

- *Stating your position* - making sure the other person knows your viewpoint.

- *Listening* - active listening means not making assumptions about what the other person is saying, or what they really mean. You need to be patient, question (intelligently!) and be supportive. Even low-time pilots have opinions!

- *Resolving differences* - almost always, the best way to o this is ensure that the results are best for everyone concerned.

- *Providing feedback*

Factors Affecting Judgement

"Judgement is the process of recognising and analysing available information about yourself, your aircraft and the environment you are in, followed by the rational evaluation of alternatives to implement timely decisions which maximise safety. This involves the ability to evaluate risks based on knowledge, skill and experience."

In short, the process of choosing between alternatives for the safest outcome. Factors that influence the exercise of good judgement include:

- *Lack of vigilance* - vigilance (that is, keeping an eye on what's going on) is the basis of situational awareness. You need to keep a constant watch on all that is going on around you, however tempting it may be to switch off for a while on a long navex. Monitor the fuel gauges, check for traffic and engine-off landing sites, *all the time.*

- *Distraction* - anything that stops you noticing a problem, for example, slowly backing into trees while releasing a cargo net. Keep pulling back from the situation to reaffirm you awareness of the big picture.

- *Peer Pressure* - we all like to be liked, whether by people inside or outside your own company. Do they want you to fly overweight? Or fly in darkness, even though they are late back? Being too keen to please is part of a self-esteem problem, another aspect of allowing yourself to be put upon.

- *Insufficient Knowledge* - although you can look the regulations up in a book, this is not always the most convenient solution, so you need a working knowledge of what they contain, including checklists and limitations from the flight manual, etc. We don't all have the luxury of an aircraft library, or have the time to refer to it if there was one.

- *Unawareness of Consequences* - this is an aspect of insufficient knowledge, above. What are the consequences of what you propose to do? Have you thought things out thoroughly?.

- *Forgetfulness of Consequences* - similar to the above.

- *Ignoring the Consequences* - again, similar to the above, but more of a deliberate act, since you are aware of the consequences of your proposed actions, but choose to ignore them.

- *Overconfidence* - this breeds carelessness, and a reluctance to pay attention to detail or be vigilant. Also, it inclines you to be hasty, and not consider all the options available to you. This is where a little self-knowledge and humility is a great help.

Human Error
This can be present at four levels – unsafe acts (errors & violations), predispositions to unsafe acts, unsafe or inadequate supervision and organisational influences.

Habits
These are part of our lives; many are comforting and part of a reassuring routine that keeps us mentally the right way up. Others, however, are ones we could well do without, but the trouble is that they can be very difficult to break, because the person trying to break them is the very person trapped by them. We learn habits as children, simply in order to survive. Despite our true nature, we quickly find out that if we want food, attention or "strokes", as the Americans say, we have to behave in certain ways, depending on the nature of our parents; in some

families getting noticed demands entirely different behaviour than in others, mostly opposite to what we really are, which is one source of stress. In certain circumstances, habits can be dangerous - if you can't do anything about them, we need at least to be aware of them.

Training is all very well, but don't let it limit your thinking. Also, don't confuse *stereotyping* with *probability*. You can always accept a probability that certain actions will solve a similar problem to one you've had before, but stereotyping implies that the same actions work every time.

Attitudes
Flying requires considerable use of the brain, with observation and/or reaction to events, both inside and outside the aircraft. Psychology and aviation have been used to each other for some time; you may be familiar with selection tests and interviews. Part of why accidents happen is that some people are accidents waiting to happen! This depends on personality, amongst other things, and we will look at this shortly. However, personality is not the only factor to be aware of on the flight deck. *Status*, *Role* and *Ability* are also important. Having two Captains on board, with neither sure of who's in charge can be a real problem! Either they will be scoring points off each other, or be too gentlemanly, allowing an accident to happen while each says "after you". How do you sort out the mess if you have someone in the left seat who is a First Officer pretending to be a Captain, and someone in the other seat who is a Captain pretending to be a First Officer?

What type of person is a pilot?

Having decided what product we are selling (safe arrival), we can now talk about the best kind of person to produce it. We certainly have more intelligence than the average car driver. Or do we? Passing exams doesn't mean you're capable of doing a decent job or handling a crisis. There are stupid solicitors, professors, you name it. I have flown with 17,000-hour pilots who I wouldn't trust with a pram, and 1,000-hour types with whom I would trust anything.

I think it's fair to say that the public typically think of pilots (when they think of them at all) as outgoing types, often in the bar and having a lark, an image that has come about from all those World War II movies. To be fair, if you were cold, hungry, tired, frightened and inexperienced, you would probably behave that way, too, but life today is quite different. I think a pilot should be a synthesis of the following headings:

- *Meticulous* - being prepared to do the same thing, the same way, *every time*, and not get bored, because that's the way you miss things.

- *Forward Thinking* - in just the same way that the advanced driver is ready to deal with a corner before going into it, the advanced pilot knows that the load underneath will carry on if the helicopter slows down, and positions the controls as best he can. Unfortunately, this ability only comes with experience, but it's never too late to start.

- *Responsible* - the "responsible position" that you hold as a

commander is one in which you act with minimum direction but are personally responsible for the outcome of your activities. In other words, you are responsible for the machine without being directed by any other person in it.

- *Trustworthy* - people must be able to *trust* you – all of aviation runs on it. You trust the previous pilot not to have overstressed the machine, or to really have done 4.3 hours and not 6. Signatures count for a lot, and, by extension, your word.

What is the common thread that unites all competent people?

- *Intelligence*

- *Personality.* This can be defined as "The sum total of the physical, mental, emotional and social characteristics of an individual". Generally, to be accident prone, you are either under- or overconfident. With the former, situations will tend not to be handled properly, and with the latter, situations not appreciated properly. You might also be aggressive, independent, a risk taker, anxious, impersonal, competitive, invulnerable and have a low stress tolerance, which, when you think about it, are all based on attention-seeking and fear. However, the real area where personality comes to the fore is during interactions with other people; behaviour tends to breed behaviour. Crews are frightened to deal with the

Captain, and Captains won't deal with crews.

- *Leadership is teamwork.* Leadership has been defined as facilitating the movement of a team toward the accomplishment of a task, in this case, the crew and the safe arrival of their passengers. This is a better definition than "Getting somebody to do what you want them to do" which implies a certain amount of manipulation, something more in the realm of management as a scientific process. A Leader, as opposed to a Manager, is a more positive force, inspirational, nurturing and many other words you could probably think of yourself.

- *Personal qualities* to passengers and colleagues.

On top of *personality traits*, which you are born with, the accident-prone person also has undesirable *attitudes*, which are acquired. These can include:

- *Impulsivity.* Doing things without forethought – such people don't stop to think about what they're doing. Not so fast! Apply your training! The opposite is *indecisive.*

- *Antiauthority.* These people don't like being told what to do. They may either not respect the source of the authority, or are just plain ornery (with a deep source of bottled-up anger). Very often there's nothing wrong with this - if more people had questioned authority, we wouldn't have had half the wars,

or we wouldn't get passengers pressurising pilots to do what they shouldn't. However, regulations have a purpose. They allow us to act with very little information, since everything is supposed to be predictable, although that doesn't mean that rules should blindly be obeyed - sometimes breaking the rules saves lives - the DC10 that had an engine fall off during takeoff could have kept flying if the nose had been lowered a little for speed, instead of being set at the "standard" angle of 6°, as taught on the simulator, which, in this case, stalled the aeroplane (now there's a little red light in the cockpit that tells you when an engine falls off). The opposite is *brainwashed.*

- *Invulnerability.* People like this think that nothing untoward can happen to them, so they take more risks, or push the envelope – humility is the antidote. Repetitive tasks must be done as if they were new every time, no matter how tedious - you can guarantee the one time you don't check for water in a fuel drum, it will be there! The opposite is *paranoid.*

- *Macho* people are afraid of looking small and are always subject to peer pressure, which means they care a lot about what other people think of them, leading to the idea that they have a very low opinion of themselves. Thus, they take unnecessary chances for different reasons than so-called Invulnerable people, above.

These are typically the high-powered intimidating company executives who have houses in the middle of nowhere with no navaids within miles of the place. Such people may subconsciously put themselves in situations where they push the weather to test their own nerve. You have to learn to stick up for yourself, with management and passengers. The opposite is *wimp*.

* *Resignation*. The opposite of impulsive. The thought that Allah will provide is all very well, but the Lord only helps those who help themselves – you've got to do your bit too! If you want the Lord to help you win the lottery, you have to buy the ticket first! The opposite is *compulsive*.

As you can see when you compare the opposites, each side of each coin above is as bad as the other – we should be somewhere in the middle, with a possible slight bias towards antiauthority.

One way of controlling hazardous personality traits is to keep a tight hold on the factors in this mnemonic:

Stress

Weather

Exposure To Risk

Aircraft

Time Constraints

Risk Management

One definition of *risk* is the chance that a situation, or the consequences of one, will be hazardous enough to cause harm, injury or loss.

To have absolutely no risk, of course, we shouldn't take off at all, but we obviously can't do that, so we have to have some method of evaluating risk against some sort of yardstick in order to get the job done. *Risk management* is the key, best used in an ample-time decision-making situation.

For example, in a helicopter, it can sometimes be more dangerous to avoid the height/velocity curve (say when coming out of a confined area) than to be inside it for a few seconds.

Risk management is a decision-making tool that can be applied to either eliminate risk, or reduce it to an acceptable level. Things that stop you eliminating risk entirely would either be impracticality, or money.

With it, you have to first identify a hazard, analyse any associated risks, make a decision and implement it (with a *risk strategy*) and monitor the results, with a view to changing things if need be. However, this depends on the *perception* of a risk, and the difference between yours, the management's and the customer's can be quite startling. Outside influences include weather, traffic and obstacles. Internal ones can be maintenance, fatigue, or the culture of the company.

Risk is equal to probability multiplied by the consequences of what you propose to do, and your exposure. You essentially have four choices – you can either not do the job, mitigate the effects of the risk, transfer it (buy insurance) or eat it.

You could always try and prehandle situations – that is, make as many decisions as possible ahead of time, as part of your flight planning – most important, though is to *leave yourself a way out*. For example, always be aware, when dropping water, that you may have to get out of a hot hole with the load on – don't assume that the bucket will work and you will be light enough to escape! Is the weather closing in behind you? Have you gone into a confined area to far forward and boxed yourself in?

Personal Checklist

Use this as part of your personal checklist before getting airborne.

Illness – degrades your abilities

Attitude – be professional

Medication – no self-prescription

Stress – know yourself honestly

Alcohol – bottle to throttle times

Fatigue – be rested and unhurried

Eating – ensure regular meal

Glossary

These are mainly from the ANO and ICAO documentation, but stuff from other countries is included where it makes the point clearer.

AD
A eronautical Directive. An instruction issued by a competent authority that requires an action to ensure that an aeronautical product conforms to its type design and is safe for flight. In other words, an AD is a notice that requires a mandatory maintenance action to be carried out before the aircraft flies again. It should more properly be called an *A irworthiness Directive.* If an AD is coming up, you must do all your planned flying within the due time, as you may not go over it. However, if you have reasonably planned a flight, but are going to go over the allotted time, you can complete the flight once started, with no stops. If you have to stop for fuel, you must remain there until the AD is satisfied.

Aerial work
A commercial air service, but not for air transport or flight training. See article 129.

Aerial work aircraft
An aircraft (not public transport) flying for aerial work.

Aerial work undertaking
An undertaking whose business includes doing aerial work.

Aerobatic manoeuvres
Include loops, spins, rolls, bunts, stall turns, inverted flying, etc., usually with a change of bank angle greater than 60°, an abnormal attitude or acceleration not incidental to normal flying.

Aerial sightseeing flight
One carried out as part of a sightseeing operation or any other commercial flight conducted for sightseeing from the air.

Aerodrome

Any area of land or water designed, equipped, set apart or commonly used for the landing and departure of aircraft (including vertically), assuming it has not been abandoned.

Aerodrome control service

An air traffic control service for aircraft on the manoeuvring area or apron of an aerodrome where the service is being provided, or which is flying in, or near, the aerodrome traffic zone by visual reference to the surface.

Aerodrome operating minima

The cloud ceiling and RVR for take-off, and the DH or MDH, RVR and visual reference for landing, which are minimum acceptable for the operation of that aircraft at that aerodrome.

Aerodrome traffic

That on the movement area of an aerodrome and aircraft at or near it.

Aerodrome traffic zone

A 2 nm circular airspace round an aerodrome notified for rule 39, unless within that of a controlling aerodrome, with the longest runway less than 1850 metres, from the surface to 2000 ft agl or msl when offshore. If the longest runway is greater than 1850 metres, the radius of the circle becomes 2.5 nm. Offshore, it is 1.5 nm.

Aeronautical beacon

An aeronautical ground light visible continuously or intermittently to designate a particular point on the surface of the earth.

Aeronautical ground light

A light provided as an aid to air navigation, other than one on an aircraft.

Aeronautical radio station

A radio station on the surface, transmitting or receiving signals to assist aircraft.

Aeroplane

According to ICAO, "a power driven heavier-than-air aircraft, deriving its lift in flight chiefly from aerodynamic reactions on surfaces which remain fixed under given conditions of flight".

AGL

Above Ground Level.

Air carrier

Any person operating a commercial air service.

Aircraft

Any machine that can derive support in the atmosphere from reactions of the air, other than one designed to derive support from reactions against the earth's surface of air expelled from it, including a rocket.

Aircraft identification plate

A fireproof plate on an aircraft that identifies it as a whole.

Aircraft Stand

Part of an apron for parking aircraft.

Air operator

The holder of an Air Operator Certificate.

Air operator certificate
A certificate issued by the Authorities that authorizes its holder to operate a commercial air service.

Airship
A power-driven, lighter-than-air aircraft.

Air show
An aerial display or demonstration before an invited assembly of people by one or more aircraft.

Air time
For technical records, the time from when an aircraft leaves the surface until it comes into contact with it again at the next point of landing.

Air traffic control unit
A person appointed by the CAA or anyone else maintaining an aerodrome or place, to give instructions, advice or information by radio to aircraft in the interests of safety, but not including flight information service officers.

Air transport undertaking
One whose business includes flights for public transport of passengers or cargo.

Airworthy
With regard to an aeronautical product, in a fit and safe state for flight, as per its type design.

Alternate aerodrome
One to which a flight may proceed when landing at the destination is inadvisable or impossible.

Altitude
The vertical distance of a level or a point, measured from mean sea level. It is represented by the QNH.

Annual costs
An estimate for the year beginning the previous 1st of January, of the costs of keeping, maintaining and operating the aircraft, excluding direct costs and those incurred without a view to profit.

Annual flying hours
An estimate of the hours flown, or to be flown, by an aircraft from the previous 1st of January.

Approach control service
An ATC service for aircraft not receiving an aerodrome control service, that is nevertheless flying in, or near an aerodrome traffic zone, whether or not with visual reference.

Approach to landing
That portion of a flight below 1000 ft above the relevant DH or MDH.

Appropriate aeronautical radio station
One serving the area in which an aircraft is for the time being.

Appropriate air traffic control unit
Either that serving the area where an aircraft is, or the area it intends to enter and with which it must communicate before entering.

Apron
Part of a land aerodrome for accommodating aircraft for loading and unloading passengers, cargo, fuelling, parking or maintenance.

APU

A uxiliary power unit. Any power unit delivering rotating shaft power or compressed air, or both, not intended for direct propulsion of an aircraft.

Area control centre

An air traffic control unit providing an area control service to aircraft within a notified flight information region, not receiving an aerodrome control or approach control service.

Area control service

An air traffic control service for aircraft not flying in or near an aerodrome traffic zone except for one notified as being subject to an area control service.

Area navigation equipment

Equipment on an aircraft enabling it to navigate on any desired flight path in the coverage of appropriate ground-based navigation aids, or within the limits of that equipment, or a combination of the two.

ASL

Above Sea Level.

ATC clearance

An authorization from ATC authorizing an aircraft to proceed within controlled airspace under conditions specified. If it is not suitable, you may request and, if practicable, obtain an amended one.

ATC instruction

A directive from an ATC unit for ATC purposes.

Authorised person

Any constable, or person authorised by the Secretary of State (article 118), or the CAA. That is, one given authority by the CAA to perform certain functions on its behalf. Paragraph 15 of Schedule 1 to the Civil Aviation Act 1982 permits the CAA to authorise any member or employee of it to do so. Because of the constraints of subordinate legislation, the activities of Authorised Persons must necessarily be restricted to those within the CAA's responsibility, that is, those in Section 3 of the enabling Act.

CAA Resolution 21, dated 5 June 1975, notes that a quorum for authorising anyone to perform functions on its behalf is one member (a quorum is a minimum number of people). Therefore, a minimum of one member of the Board of the CAA must form a Board Meeting to do this. Thus, the appointment of an Authorised Person must be made by at least one person authorised to authorise, so to speak, be it a Board Member or somebody duly delegated. If the card carried by the AP does not carry any proper indication that the authorised was in fact so authorised then it may not be valid (this may sound like a cheap point, but as the CAA is not a natural person in its own right, its range of action is limited without proper procedures).

To enable it to be produced in Court as part of a case, that is, to be *admissible evidence*, a document must be properly authenticated. Paragraphs 16 and 17 of Schedule 1 (of the 1982 Act) provide that a document received in evidence should have the seal of the CAA on it, and that the seal itself is not valid unless authenticated by the signature of the Secretary of the CAA (or somebody duly authorised by the Board). In the light of this, unless a document carried by an alleged AP has such a seal and signature on it, and is dated because it is subordinate legislation, there may be no proof (acceptable to a court, anyway) that he is in fact an AP, and therefore probably should not have wasted your time asking all those questions in the beginning.

A constable is an Authorised Person, but don't forget to ask for his warrant card or note his collar number (*every* policeman has one, including detectives, so don't let them tell you otherwise). A policeman *in full uniform* is properly appointed, but if he's not wearing a hat, or his buttons are undone, his authority is in question (a motorist was stopped by two policemen without hats, and the charges were thrown out). He may be from plain clothes, of course, but "constable" in the normal sense doesn't usually mean "detective".

Balloon
A non-power-driven lighter-than-air aircraft.

Captive balloon
One which, when in flight, is attached by a restraining device to the surface.

Captive flight
Flight by an uncontrollable balloon which is attached to the surface by a restraining device.

Cargo
Includes mail and animals

Cargo aircraft
An aircraft that is not a passenger aircraft, carrying goods or property.

Category
With reference to flight crew licensing, the classification of aircraft as aeroplanes, balloons, gliders, gyroplanes, helicopters or ultra-lights. Otherwise, a grouping of aircraft based on use or limitations, such as normal, utility, aerobatic, commuter and transport.

Certificate of Airworthiness
Includes any validation and flight manual, performance schedule or other document referred to.

Certificate of maintenance review
See article 10(1)

Certificate of release to service issued under JAR-145
Means just that.

Certificate of release to service issued under the Order
Se article 12(7)

Certificated for single pilot operation
Refers to an aircraft not required to carry more than one pilot because of its current C of A, that last in force, that of an identical aircraft, or a permit to fly.

Chief officer of police for any area of the UK
For Scotland, the Chief Constable for any area. For Northern Ireland, the Chief Constable of the RUC.

Child restraint system
A device, not a safety-belt, designed to restrain, seat or position a person defined as a child.

Class
Machines with similar operating characteristics, that is, single-engined, multi-engined, centre-line thrust, land or sea aeroplanes.

Class *X* airspace
Airspace respectively notified as such, where *X* refers to the classification.

Class rating
For aeroplanes, see paragraph 1.220 of JAR-FCL 1.

Cloud ceiling
The vertical distance from the elevation of an aerodrome to the lowest part of any cloud visible from it sufficient to obscure more than one-half of the sky.

Commander
The member of the flight crew so designated by the operator, or whoever is for the time being the pilot in command. In other words, the person who gets it in the neck at the subsequent Board of Inquiry.

Commercial air service
Any use of aircraft for hire or reward.

Commonwealth
UK, Channel Islands, Isle of Man, countries in Schedule 3 to the British Nationality Act 1981, and all territories forming part of HM's dominions or where HM has jurisdiction.

Competent authority
In relation to UK, the CAA, and to any other country authority responsible under its law for promoting the safety of civil aviation.

Conditional sale agreement
See section 189 of the Consumer Credit Act 1974

Congested area
For a city, town or settlement, any area substantially used for residential, industrial, commercial or recreational purposes (including a golf course).

Contracting State
Any State (including UK) party to the Chicago Convention.

Controllable balloon
A balloon, not a small one, capable of free controlled flight.

Controlled airspace
That notified as Class A, Class B, Class C, Class D or Class E.

Control area
Controlled airspace further notified as a control area upwards from a notified altitude or flight level.

Control zone
Controlled airspace further notified as a control zone extending upwards from the surface.

Convention
The *Convention on International Civil Aviation* signed at Chicago in 1944, as amended.

Co-pilot
A pilot subject to the direction of another one in the aircraft.

Country
Includes a territory.

Crew
A member of the flight crew, or someone on the flight deck, appointed to give or to supervise the training, experience, practice and tests required under article 34(3), or a cabin crew member.

Crew member
A person assigned to duty in an aircraft during flight time.

Critical engine
One whose failure most adversely affects performance or handling of an aircraft.

Critical surfaces
The wings, control surfaces, rotors, propellers, horizontal stabilizers, vertical stabilizers or other stabilizing surface of an aircraft and, for one with rear-mounted engines, includes the upper surface of its fuselage.

Danger area
Airspace notified as such within which activities dangerous to flight may take place as and when notified.

Dangerous goods
Anything that poses a risk to life, property or the environment, such as aerosols, solvents, paints, chainsaws, matches, stoves, car batteries, gas tanks and even perfume under the right circumstances.

Day
The time from half an hour before sunrise until half an hour after sunset (both times exclusive), sunset and sunrise being determined at surface level. In some countries, the definition also means where the centre of the sun's disc is less than 6° below the horizon (just in case you're up in the Frozen North, where the Sun doesn't really set).

Direct costs
For a flight, the costs actually and necessarily incurred in connection with it without a view to profit but excluding anything paid to the pilot.

Director
See section 53(1) of the Companies Act 1989.

Ditching
The forced landing of an aircraft in water.

EEA Agreement
The Agreement on the European Economic Area signed at Oporto on 2nd May 1992 as adjusted by the Protocol signed at Brussels on 17th March 1993.

EEA State
A State which is a contracting party to the EEA Agreement.

Elementary work
Maintenance tasks that may be completed by an unqualified person.

Empty weight
The total weight of the airframe, engines, fixed ballast, unusable fuel, the maximum amount of normal operating fluids, such as oil, power plant coolant, hydraulic fluid, de-

icing and anti-icing fluid, not including potable water, lavatory pre-charge fluid or that for injection into the engines, and all installed equipment.

Evaporation

The conversion of liquid to vapour at temperatures below its boiling point (vapourisation occurs at the boiling point).

Farmer

A person whose primary source of income is derived from tilling the soil, raising livestock or poultry, dairy farming, the growing of grain, fruit, vegetables or tobacco, or any other operation of a similar nature.

Federal Aviation Regulations

As published by the USA.

Fixed

About a light, means having a constant luminous intensity when observed from a fixed point.

FL

Flight Level. Officially, one of a series of levels of equal atmospheric pressure, separated by notified intervals and expressed as hundreds of feet indicated at that level on an ISA pressure altimeter. In normal usage, altitude in hundreds of feet, on an altimeter set to 29.92 inches of mercury or 1013.2 mb.

Flammable

A quality of a substance that means it will burn. This word replaced *inflammable* in the 60s.

Flash point

The lowest temperature of a liquid at which flammable vapour is given off.

Flight attendant

A crew member assigned duties in the interest of passengers.

Flight authority

A C of A or permit to fly showing an aircraft's fitness for flight under Article 31 of the Convention.

Flight crew

Members of the crew of an aircraft who undertake to act as pilot, flight navigator, flight engineer and flight radiotelephony operator, which means that when you're in the pub, you are not flight crew, as you are not undertaking to act as such.

Flight deck duty time

Time spent by a crew member at a relevant position during flight time.

Flight duty time

Time starting when a crew member reports for a flight, or standby, and finishes at engines off or rotors stopped at an appropriate period after the end of the final flight.

Flight information service unit

A person appointed by the CAA or any other person maintaining an aerodrome to give information by radio to aircraft flying or intending to fly within the aerodrome traffic zone, and grant or refuse permission, under Rule 35 or 36(2). For an area control centre, the former applies.

Flight level

A surface of constant atmospheric pressure related to a specific pressure datum (normally 1013.2 mb or 29.92"), separated from other levels by prescribed intervals.

Flight line
A predetermined directional line of flight within a flying display area. It must be marked and clearly visible from the air.

Flight plan
Such information as may be notified for an air traffic control service unit being information provided or to be provided to that unit, relative to an intended flight or portion of a flight of an aircraft;

Flight recording system
A system with a flight data recorder, cockpit voice recorder, or both.

Flight simulator
Apparatus which simulates flight conditions on the ground.

Flight time
From when an aircraft first moves under its own power to take off until it comes to rest at the end of the flight.

Flight training
A program of ground instruction and airborne training under the flight instructor guide and flight training manual for the aircraft used.

Flight visibility
The visibility forward from the flight deck when in flight.

Flying display
Flying activity deliberately performed for an exhibition or entertainment at an advertised event open to the public.

Flying machine
For the exams, "a heavier than air power driven aircraft", as ICAO only define an aeroplane. It may be a glider, aeroplane or rotorcraft.

Formation flight
When an aircraft is flown solely with reference to another aircraft.

Free balloon
One which, which, when in flight, is not attached by a restraining device to the surface.

Free controlled flight
Where a balloon is not attached to the surface by a restraining device (other than a tether less than 5 metres long used during takeoff), and during which height is controllable by a device attached to the balloon operated by the commander or by remote control.

General lighthouse authority
See section 193 of the *Merchant Shipping Act 1995*.

Glider
A non-power-driven heavier-than-air aircraft deriving lift from aerodynamic reactions on fixed surfaces.

Goods
Anything that may be taken or placed on an aircraft as personal belongings, baggage or cargo.

Government aerodrome
An aerodrome in UK occupied by a Govt Department or visiting force.

Granted under JAR-FCL
Granted by an authority being a Full Member of JAA under a procedure assessed as satisfactory after an inspection by a licensing and a medical standardisation team.

Ground station
A location with radio transmitting and receiving equipment capable of two-way voice communications with aircraft.

Gyroplane
A heavier-than-air aircraft deriving lift from aerodynamic reactions on one or more non-power-driven rotors on substantially vertical axes.

Hang glider
One designed to carry less than two people with a typical launch weight of 45 kg (99.2 pounds) or less.

Heading
The direction in which the longitudinal axis of an aircraft is pointed, usually in true, magnetic or grid degrees from North.

Heavier-than-air aircraft
One supported in the atmosphere by lift derived from aerodynamic forces.

Height
The vertical distance of a level or a point measured from a specific datum, such as airfield datum. It is referenced by the QFE.

Helicopter
A power-driven heavier-than-air aircraft deriving lift from aerodynamic reactions on one or more power-driven rotors on substantially vertical axes.

Heliport
An aerodrome used for the arrival, landing, take-off or departure of aircraft capable of vertical take-off and landing.

High seas
Any body of water, or frozen surface, not within territorial waters of any state.

Hire or reward
Any payment, consideration, gratuity or benefit, directly or indirectly charged, demanded, received or collected by any person for the use of an aircraft.

Hire-purchase agreement
See section 189 of the *Consumer Credit Act 1974*.

Identification plate
A fireproof plate (preferably metal) with identification information.

IFR
Instrument Flight Rules. To be observed when flying in IMC.

IFR aircraft
One operating in IFR flight.

IFR flight
One conducted under IFR.

IMC
Instrument Meteorological Conditions. Those less than the minima for VMC, in terms of visibility and distance from cloud.

Infant
A person under two years of age (this may vary).

Instructor's rating

A flying instructor's rating, assistant flying instructor's rating, flight instructor rating (aeroplane) or (helicopter), type rating instructor rating (multi-pilot aeroplane) or (helicopter), a class rating instructor rating (single pilot aeroplane), an instrument rating instructor rating (aeroplane) or (helicopter).

Instrument Flight Rules

Those for flight solely with reference to instruments, as per the Rules of the Air.

Instrument Meteorological Conditions

Weather stopping flight under Visual Flight Rules.

International headquarters

As designated by Order in Council under section 1 of the International Headquarters and Defence Organisations Act 1964.

JAA

Joint Aviation Authorities, an associated body of the European Civil Aviation Conference.

JAA Full Member State

A State being a full member of the JAA.

JAA licence

A licence granted under JAR-FCL.

JAR

A Joint Aviation Requirement of the JAA with a number or letters. Reference to a numbered or lettered JAR is one to such a requirement as adopted by JAA or, where a JAR has been annexed to the Technical Harmonisation Regulation, in the form it has been thus annexed and has effect under that Regulation.

JAR-FCL licence

A licence included in Section 2 of Part A of Schedule 8.

Land

As a verb, includes alighting on water.

Land aircraft

One not capable of normal operations on water.

Landing

For other than airships, coming into contact with a supporting surface, including actions immediately before and after. Also, the act of bringing an airship under restraint, plus any acts immediately before and after.

Large Rocket

One with a total impulse from the engines over 10,240 Newton-seconds.

Latent heat

Converting water (for example) from one state to the other requires energy, which originally comes from the Sun's rays as it is evaporated in the first place, and is stored with the vapour. While there, it is known as latent, and released when the water condenses. Latent heat becomes involved when you change the form of a substance without changing its temperature.

Launch weight

The total weight of a hang glider or an ultra-light when ready for flight, including equipment, instruments, fuel or oil, but not occupants, float equipment, etc.

LDA
Landing Distance Available. The length of a runway declared available and suitable for the ground run of landing aeroplanes.

Legal personal representative
The executor, administrator, or other representative, of a deceased person.

Licence
Includes certificates of competency or validity issued with it, or required to be held in connection with it (e.g. a medical) by the law of the country it is granted.

Licence for public use
See article 103(3).

Licensed aerodrome
One licensed under the ANO.

Lifejacket
Includes any device designed to support a person individually in or on the water.

Life-limited
Parts or a part that, as a condition of the type certificate, may not exceed a specified time, or number of cycles, in service.

Lighter-than-air aircraft
One supported in the atmosphere by its own buoyancy.

Light turbulence
That which momentarily causes slight, erratic changes in altitude or attitude, or which causes slight, rapid and somewhat rhythmic bumpiness without appreciable changes in altitude or attitude.

Log book
Includes records kept either in a book, or other approved means.

Maintenance
The overhaul, repair, required inspection or modification, or removal and installation of components of, an aeronautical product, not including elementary work or servicing.

Maintenance schedule
One required for inspections and other maintenance of aircraft.

Major modification
An alteration to the type design of an aeronautical product with more than a negligible effect on weight and centre-of-gravity limits, structural strength, performance, power plant operation, flight characteristics or other qualities affecting airworthiness or environmental characteristics.

Major repair
A repair to an aeronautical product with a type certificate that causes it to deviate from the design defined by the type certificate, with more than a negligible effect on the weight and centre-of-gravity limits, structural strength, performance, power plant operation, flight characteristics or other qualities affecting airworthiness or environmental characteristics.

Manoeuvring area
That part of an aerodrome used takeoff and landing of aircraft and movement on the surface, excluding the apron and any part used for maintenance.

Marker

An object displayed agl for indicating an obstacle or obstruction, or delineating a boundary.

Marking

A symbol or group of symbols displayed on the surface of a movement area for conveying aeronautical information.

Maximum total weight authorised

The maximum total weight of an aircraft and contents, at which it may take off anywhere, in the most favourable circumstances under the C of A.

Medical attendant

A person carried to attend to any person in need of medical attention, or to be available to attend to them.

MEL

Minimum Equipment List. A document that authorizes inoperative equipment, or equipment that must be working.

Microlight aeroplane

An aeroplane with a maximum total weight of 390 kg or below (450 for a two-seater, 330 for a single seat amphibian or floatplane, and 495 for a two-seat amphibian or floatplane), and either an associated wing loading of up to 25 kg per square metre or a stalling speed of 35 kts CAS at MAUW, and designed to carry up to two people.

Military aircraft

The naval, military or air force aircraft of any country, and any being constructed for them under a contract entered into by the Secretary of State, or for which there is a certificate that the aircraft is to be treated as such.

Military Rocket

The naval, military or air force rockets of any country, and any being constructed for them under a contract entered into by the Secretary of State, or for which there is a certificate that the aircraft is to be treated as such.

Minimum Fuel

A situation where your fuel state is such that little or no delay can be accepted.

Model aircraft

One with a total weight typically under 35 kg (77.2 lbs), mechanically driven or launched into flight for recreational purposes, not designed to carry people or other living creatures.

Model rocket

One with model rocket engines that will not generate a total impulse over 80 newton-seconds, under 500kg and with a parachute or other device capable of retarding its descent.

Movement area

Part of an aerodrome for the surface movement of aircraft, including the manoeuvring area and aprons.

Nationality mark

Symbols, letters or numerals, or combinations, used by states to indicate nationality of aircraft registered in them.

Nautical mile

The International Nautical Mile, that is, 1852 metres.

Night

The time from half an hour after sunset until half an hour before sunrise (both times inclusive), sunset and sunrise being determined at surface level. In some countries, also where the centre of the Sun's disc is more than 6° below the horizon.

The ICAO definition is "between the end of evening civil twilight and the beginning of morning civil twilight, or as may be prescribed by the appropriate authority."

Non-piloted aircraft

A power-driven aircraft, other than a model, without a crew member.

Non-revenue flight

A flight that a PPL holder (aeroplanes, helicopters or gliders) may undertake under paragraph (2)(a) and (b) of the privileges in Section 1 of Part A of Schedule 8.

NOTAM

A notice to airmen about aeronautical facilities, services or procedures, or any hazard affecting aviation safety, essential to personnel engaged in flight operations.

Notified

Set forth with the authority of the CAA in a document published by or under an arrangement entered into with the CAA and entitled 'United Kingdom Notam' or 'Air Pilot', for the time being in force.

Obstacle limitation surfaces

See CAP 168, Licensing of aerodromes.

Operational control

The exercise of authority over the initiation, continuation, diversion or termination of a flight.

Operator

The person at the relevant time who has the management of an aircraft.

Overhaul

A restoration process that includes the disassembly, inspection, repair or replacement of parts, reassembly, adjustment, refinishing and testing of an aeronautical product, ensuring that the product is in conformity with the tolerances in the applicable instructions for continued airworthiness.

Owner

The person with legal custody and control of an aircraft.

Parascending parachute

One towed by cable causing it to ascend.

Passenger

A person other than a member of the crew, carried in an aircraft.

Period of duty

Between the start and end of a shift in which an air traffic controller performs, or could be called upon to perform, any functions specified for a rating in the licence.

Pilot in command (PIC)

A person in charge of the piloting of an aircraft without being under the direction of any other pilot in it.

Police officer
A member of a police force or the RUC, including reserves, and special constables, all of whom must have a warrant card.

Powered glider
An aeroplane that, with the engines inoperative, behaves like a glider.

Prescribed
Prescribed by regulations made by the Secretary of State under the ANO.

Pressurised aircraft
One that can maintain in any compartment a pressure greater than the outside atmosphere.

Private flight
A flight which is not for aerial work or public transport.

Public transport
See article 130.

Radar Vectoring
The provision of aeronautical guidance to aircraft in the form of specific headings, based on radar.

Record
See section 73 of the Civil Aviation Act 1982.

Reduced vertical separation minimum airspace
Any airspace between FL 290 and 410 (inclusive) within which a vertical separation minimum of 1000 feet or 300 metres is applied.

Registered owner
The person to whom a C of R for the aircraft has been issued.

Registration mark
The combination of letters, or letters and numerals, issued to an aircraft by a state as a registration identification.

Released flight
Flight by an uncontrollable balloon not attached to the surface by a restraining device.

Relevant overseas territory
Any colony and country or place outside Her Majesty's dominions in which Her Majesty has jurisdiction.

Repair
The rectification of deficiencies in an aeronautical product, or its restoration to an airworthy condition.

Replica
An aircraft of any scale that is a duplicate of an original military aircraft.

Replacement
Removal and replacement of a part (whether the same or not), and whether or not work is done on it, but not including parts designed to be removable solely for the checking another part, or loading cargo.

Required visual reference
That portion of the approach area of the runway, or those visual aids that, when viewed by the pilot of an aircraft, enable the pilot to assess the aircraft position and rate of change of position, to continue and complete a landing.

Restricted airspace
Airspace of fixed dimensions within which flight is restricted.

Rocket

A device propelled by the ejection of expanding gases from self-contained propellant, not dependent on the intake of outside substances, including anything designed to drop off in flight.

Rules of the Air

See article 84(1).

Runway visual range (RVR)

The distance in the direction of take-off or landing over which runway lights or surface markings may be seen from the touchdown zone by human observation or instruments near the touchdown zone.

Safety belt

A personal restraint system with either a lap strap or a lap strap combined with a shoulder harness.

Safety pilot

One who acts as a lookout for another pilot operating an aircraft in simulated instrument flight.

Saturated Vapour

That in contact with its liquid at the same temperature. Vapours used in refrigeration systems are saturated.

Saturation Temperature

The boiling or vapourisation point of a liquid. The boiling point can change according to the pressure, for example, water boils at a lower temperature when at altitude - in fact, it doesn't even have to be hot in some circumstances. Thus, if you want to stop a liquid boiling, subject it to high pressure.

Scheduled maintenance

That performed at predetermined intervals, a maintenance schedule or an AD.

Seaplane

See section 97 of the Civil Aviation Act 1982.

Sector

Part of the airspace controlled from an area control centre or other place.

Self Launching Motor Glider (SLMG)

Something like a non power-driven glider with one or more power units, that takes off under its own power.

Sensible Heat

As opposed to latent heat, that which changes the temperature of a substance without changing its form.

Serviceable

Fit and safe for flight.

Servicing

Cleaning, lubricating and replenishment of fluids not requiring disassembly.

Shoulder harness

Any device for restraining the upper torso of a person, with a single diagonal or dual upper torso straps.

Simple Single Engine Aeroplane

For the (National) PPL(A), an aeroplane with one piston engine under 2000 kgs Max All Up Weight, not a microlight or a SLMG.

Skill test

See paragraph 1.001 of JAR-FCL 1 for aeroplanes and 2.001 in JAR-FCL 2 for helicopters. For pilots, may involve something concerning sex and booze.

SLMG
Self Launching Motor Glider.

Small aircraft
Any unmanned aircraft, other than a balloon or a kite, weighing not more than 20 kg without fuel, but including articles or equipment installed in or attached to it at the start of its flight.

Small balloon
One not exceeding 2 metres in any linear dimension, including baskets, etc. attached to it.

Small Rocket
One with a total impulse from its engine(s) not above 10,240 Newton-seconds.

Special VFR flight
A VFR flight authorized by ATC within a control zone where IFR would normally apply (see *Rules of the Air*).

Stopway
A rectangular area at the end of a runway in the direction of take-off, with the same width, prepared as a suitable area for stopping in a rejected take-off.

Suitable accommodation
A single-occupancy bedroom subject to minimal noise, well ventilated, with facilities to control temperature and light, or accommodation suitable for the site and season, with a minimal level of noise and providing adequate comfort and protection from the elements.

Superheated Vapour
That with its temperature above the boiling point of its liquid.

Surface
Ground or water, including when frozen.

State of the operator
The location of the principal place of business, or permanent residence if there is none.

Take-off
For other than airships, leaving a supporting surface, including the take-off run, plus acts just before and after. For an airship, freeing it from restraint, including acts just before and after.

Technical Harmonisation Regulation
Council Regulation (EEC) No 3922/91, on harmonisation of technical requirements and administrative procedures in civil aviation.

Telecommunications System
and public telecommunications system. See sections 4 and 9 (1) of the *Telecommunications Act 1984*.

Tethered flight
Flight by a controllable balloon within limits imposed by a restraining device.

TODA
Take-off Distance Available. The total of the *take-off run available* (below) and clearway.

TORA
Take-off Run Available. The length of a runway declared available and

suitable for the ground run during take-off.

Touring motor glider
See paragraph 1.001 of JAR-FCL 1.

Track
The projection on the surface of an aircraft's path, in true, magnetic or grid degrees from North.

True Mach number
Ratio of TAS to local speed of sound.

Type rating
For aeroplanes, see paragraph 1.215 of JAR-FCL 1. For helicopters, 2.215 of JAR-FCL 2.

Uncontrollable balloon
A balloon, not being a small one, not capable of free controlled flight.

UK licence
One included in Section 1 of Part A of Schedule 8.

UHF
Ultra High Frequency.

UK licence for which there is a JAR-FCL equivalent
PPL, CPL, ATPL (see Section 1 of Part A of Schedule 8).

UK licence for which there is no JAR-FCL equivalent
Any licence in Section 1 of Part A of Schedule 8 not mentioned above.

UK reduced vertical separation minimum airspace
UK airspace notified as such for article 48.

UTC
Universal Coordinated Universal Time. What used to be called Greenwich Mean Time.

Valuable consideration
Any right, interest, profit or benefit, forbearance, detriment, loss or responsibility accruing, given, suffered or undertaken pursuant to an agreement, of more than a nominal nature.

Vessel
Any ship, boat or other floating structure, other than an aircraft, used for navigation on water.

VFR
Visual Flight Rules.

VFR aircraft
One operating in VFR flight.

VFR flight
One conducted under visual flight rules.

VHF
Very High Frequency.

Visiting force
See the *Visiting Forces Act 1952.*

Visual Flight Rules
As prescribed by the *Rules of the Air.*

Visual Meteorological Conditions
Those permitting flight under VFR

VMC
Visual Meteorological Conditions. Weather expressed in terms of visibility, distance from cloud and ceiling equal to or better than that specified from time to time.

Index

Ground Effect, 38, 47
Ground Resonance, 37
gusts, 25, 40, 81
gyroplane, 179, 180, 218
gyroplanes, 265
gyroscope, 31, 71
gyroscopic precession, 25

H

Hail, 95, 105
hang gliders, 195
harnesses, 95
head-on, 187
Height/Velocity Curve, 167
helicopter, 35, 179, 180, 183, 184,
   217, 218, 219, 220, 271
helicopters, 2, 3, 4, 10, 28, 36, 37,
   38, 40, 42, 47, 51, 52, 53, 54, 58,
   59, 60, 61, 62, 63, 83, 84, 86,
   116, 118, 154, 158, 162, 167,
   180, 181, 198, 215, 221, 227,
   265, 274, 276, 278
Her Majesty's Government, 213,
   223
high blood pressure, 251
high seas, 270
hire or reward, 266, 270
hoar frost, 98
Holding, 40
horizontal stabiliser, 8
Hydraulics, 59
Hydroplaning, 164, 165
Hypoxia, 251

I

IAS, 56, 67
IATA, 212
ICAO, 85, 105, 106, 199, 206, 207,
   212
Ice, 47, 97
icing, 47, 98, 99
identification plate, 196, 262, 270
IFR, 85, 180, 217, 218
ILS, 125, 127
IMC, 209

immunisation, 252
incident, 208
indelible pencil, 223
induced drag, 24, 30, 38
Induced Drag, 23
induced power, 39
industrial pollutants, 90
Inlet Guide Vanes, 53
inlet manifold, 45, 48
INS, 132
insolation, 89
inspections, 197, 216
instruction, 179, 180, 183, 195,
   213, 214, 217, 218, 225
instructor, 179, 180, 183, 195, 213,
   214, 218, 271
insurance, 207
Interference Drag, 24
International Law, 212
inversion, 79, 89, 90, 91, 92, 94, 97
inversions, 83, 90
inverter, 120
ionosphere, 114, 122, 132
IR, 179, 180, 181, 218
is *Idle Cut Off*, 48
ISA, 88, 204
Isle of Man, 179, 217, 266
isobars, 79, 80, 88, 96, 102, 146
isothermal layer, 89

J

JAR, 177, 180, 182, 215, 216, 218,
   223, 265, 266, 270, 271, 276,
   278
JAR-145, 215, 223, 265
JAR-FCL 2, 276, 278
Jeppesen, 156
jet engine, 51, 54, 117
jet lag, 248
jettisoning, 220
Journey Log, 193
JSP 318, 223

K

katabatic wind, 81

Bibliography

Sources of information for this book and other suggested reading matter:

Title	Author	Publisher	Notes
Stick and Rudder	Wolfgang Langewiesche	McGraw-Hill	
Studies for Student Pilots	Michael Royce	Pitman	
The Private Helicopter Pilot's Guide	Steve Sparrow	Steve Sparrow	
The Art & Science of Flying Helicopters	Shawn Coyle	Shawn Coyle	
The Helicopter and How it Flies	John Fay	David & Charles	
Astronomy for Night Watchers	J B Sidgwick		
AIP Canada			
A Race on the Edge of Time	David E Fisher	McGraw-Hill	Darned good book about the history and development of radar in WWII.
Chickenhawk	Robert Mason		Excellent story of a helicopter pilot's training and experiences in the Vietnam war.
Positive Flying	Taylor & Guinther	Delacorte	Many useful fixed wing tips
Fit For Life	Richard and Marilyn Diamond		Food Combining
The Helicopter Pilot's Handbook	Phil Croucher	Electrocution	
Operational Flying	Phil Croucher	Electrocution	
CARs in Plain English	Phil Croucher	Electrocution	
Amateur Radio Study Guides			

By The Same Author

CARs in Plain English
Canadian Aviation Regs translated!

The ANO in Plain English
British Aviation Regulations translated!

The Helicopter Pilot's Handbook
One snag with helicoptering is that there are virtually no flying clubs, at least of the sort that exist for fixed wing, so pilots get very little chance to swap stories, unless they meet in a muddy field somewhere, waiting for their passengers. As a result, the same mistakes are being made and the same lessons learnt separately instead of being shared – it's comforting sometimes to know that you're not the only one to have inflated the floats by accident! Even when you do get into a school, there are still a couple of things they don't teach you, namely that aviation runs on paperwork, and how to get a job, including interview techniques, etc— flying the aircraft is actually a very small part of the process.

Another drawback is that nobody really *tells* you anything, either about the job you have to do (from the customer) or how to do it (the company) – because of previous supplies from the military, people are used to pilots who can turn their hand to anything and assume you know it all – you will always be up against the other guy who managed to do it last week! Sure, there will be training, but, even in the best companies, this will be relatively minimal.

This book is an attempt to correct the above problems by gathering together as much information as possible for helicopter pilots, old and new, professional and otherwise, in an attempt to explain the *why*, so that the *how* will become easier (you will be so much more useful if you know what the customer is trying to achieve). It contains all the helicopter-specific information from *Operational Flying*, which is more to do with the admin side, plus additional chapters about two popular light helicopters, the Bell 206 and AS 350.

The BIOS Companion
A book that deals with all those secret settings in your computer's BIOS, plus tons of data for troubleshooters.

> "I already had about HALF of the information, and to get THAT much, I had to get several books and web pages. GOOD JOB!!

> I had more time to go thru the book and think that you should change the word "HALF" to "FOURTH".

> I commend you on the great job you did. That's a hell of a lot of work for any major company to do, let alone an individual.

> Again, Thank you

> *Craig Stubbs*

> Thank! I really appreciated this. I read it and was able to adjust my BIOS settings so that my machine runs about twice as fast. Pretty impressive.

> Thanks again.

> --*Tony*

Available from
www.electrocution.com